Atmospheric
Explorations

Technology Press Books in Science and Engineering

Notes on Analog-Digital Conversion Techniques
Edited by Alfred K. Susskind
Lectures on Ordinary Differential Equations
By Witold Hurewicz
Atmospheric Explorations
Edited by Henry G. Houghton
Ceramic Fabrication Processes
By W. D. Kingery
Currents, Fields, and Particles
By Francis Bitter
The Theory and Technique of Ship Design
By George C. Manning
Spheroidal Wave Functions
By J. A. Stratton, P. M. Morse, L. J. Chu, J. D. C. Little, and F. J. Corbató
Principles of Electric Utility Engineering
By C. A. Powel
Dielectric Materials and Applications
By A. R. von Hippel
Transient Analysis of Alternating-Current Machinery
By Waldo V. Lyon
Physical Meteorology
By John C. Johnson
The Mathematics of Circuit Analysis
By E. A. Guillemin
Magnetic Circuits and Transformers
By Members of the Electrical Engineering Staff, M.I.T.
Applied Electronics
Second Edition by Truman S. Gray
Electric Circuits
By Members of the Electrical Engineering Staff, M.I.T.
Fatigue and Fracture of Metals: A Symposium
Edited by William M. Murray
Metadyne Statics
By Joseph Maximus Pestarini
The Prefabrication of Houses
By Burnham Kelly
Methods of Operations Research
By Philip M. Morse and George E. Kimball
The Transmission of Nerve Impulses at Neuroeffector Junctions and Peripheral Synapses
By Arturo Rosenblueth
An Index of Nomograms
Compiled and Edited by Douglas P. Adams
The Extrapolation, Interpolation, and Smoothing of Stationary Time Series
By Norbert Wiener
Index Fossils of North America
By H. W. Shimer and R. R. Shrock
Wavelength Tables
Measured and Compiled under the Direction of G. R. Harrison
Tables of Electric Dipole Moments
Compiled by L. G. Wesson
Analysis of the Four-Bar Linkage
By John A. Hrones and George L. Nelson

Edited by
Henry G. Houghton
Professor of Meteorology
Massachusetts Institute of Technology

Atmospheric Explorations

Papers of the Benjamin Franklin Memorial Symposium of the American Academy of Arts and Sciences

Published jointly by
The Technology Press
of Massachusetts Institute of Technology
and
John Wiley & Sons, Inc., New York

Chapman & Hall, Ltd., London

Library of Congress Catalog Card Number: 57-13445
Printed in the United States of America

Preface

IN CELEBRATING the two hundred and fiftieth anniversary of the birth of its late Fellow, Benjamin Franklin, the American Academy of Arts and Sciences held a Symposium on January 11, 1956, at which the papers reproduced in this volume were presented and discussed. From the many facets of Franklin's full and productive life the Academy chose to honor him through a delightful and informative account of his life in France by Professor Morris G. Bishop, a performance on a replica of Franklin's glass armonica by E. Power Biggs, and the scientific symposium.

Franklin was one of America's first and greatest scientists. His scientific observations and theories are all the more remarkable today in the light of the tremendous scientific advances of the past two centuries. Of the many scientific topics that he considered, none attracted his interest more than electricity and its manifestations in the atmosphere. He observed, and presented theories on, many other atmospheric phenomena, including "northeasters" and the aurora borealis. It was natural to select atmospheric electricity as one of the topics of the symposium. The upper atmosphere was chosen as the second subject, not only in view of Franklin's interest in the aurora, but because he would certainly have been intrigued by the often exotic phenomena of the high atmosphere that have been revealed by recent researches.

Because of a necessary preoccupation with the business of earning a living, Franklin's active interest in science did not begin until he was in his forties. His experiments with electricity were initiated with the receipt of a small electrostatic machine from his friend Peter Collinson of London. He discovered that there were two kinds of electricity, positive and negative, and correctly identified lightning as a form of electricity. His studies of the efficacy of pointed bodies in drawing off the "electrical fire" led to his invention of the lightning rod and to practical rules for the avoidance of lightning strikes.

Franklin stated that thunderclouds are most commonly negatively charged but occasionally carry positive charges. These important results have been rediscovered only in the present century by investigators equipped with complex instruments and a knowledge of modern electrical theory. This is an outstanding example of Franklin's amazing scientific intuition and his skill as an observer.

During the 1750s Franklin submitted three papers on electricity to the Royal Society in London. At first they were all rejected as not worthy of publication in the Society's Transactions. Franklin's theories were at such variance with the current ideas of electricity that it was assumed they must be erroneous. Later, the Royal Society not only published the papers but awarded Franklin its Copley Medal for his discoveries.

Franklin was particularly interested in the vortical motions of waterspouts and whirlwinds. He describes how he followed a whirlwind for some distance on horseback until it entered a dense forest, scattering small branches about him.

Engaging in another of his scientific pursuits, astronomy, Franklin was unable to observe a lunar eclipse in Philadelphia because of the clouds and rain of an extensive storm. By correspondence, he found that the eclipse had been visible in Boston which was subsequently visited by a northeaster. From this scanty information, Franklin deduced that the storm had traveled from the southwest towards the northeast, even though the wind was in the opposite direction. He computed the speed of travel of the storm as 100 miles per hour, a probable overestimate due to his lack of data from other localities. On the basis of such very limited observations, Franklin drew the remarkably accurate conclusions that such storms probably originated in the Gulf of Mexico and that they were similar in structure, though of much larger size, to the water spouts and whirlwinds he had studied at close range.

In 1779 Franklin proposed a theory of the aurora borealis. He considered that the aurora was an electrical discharge in the high atmosphere where the low pressure would result in an increased electrical conductivity. Although his ingenious explanation has proved to be incorrect, it included a meridional circulation at high levels in the proper direction.

Historians have recorded Franklin's accomplishments in science in proper context with the many and varied activities of his full and fruitful life. In paying tribute to a distinguished colleague, scientists offer their best in the form of their own contributions to science. Thus, the papers in this little volume are not historical treatises on Franklin's works and time but, rather, represent the latest and best current developments and ideas in the fields discussed. Without exception the authors have distinguished themselves in their chosen topics. No pretense is made that the coverage of the subjects is complete; this could have been accomplished only at the cost of superficiality. This is particularly true in the case of the upper atmosphere concerning which many important phenomena, including the aurora borealis, are not mentioned at all. An effort was made here to present the points of view of the physicist and the meteorologist, each of whom looks at this vast area with glasses of a different hue. In an age of specialization we must continually stress the basic unity of all the physical sciences, and this purpose is aided here by a common laboratory, the upper atmosphere.

Henry G. Houghton

Cambridge, Massachusetts
March 1958

Contents

1 Atmospheric Electricity

ROSS GUNN
Director, Office of Physical Research
U. S. Weather Bureau
Washington, D. C.

I

The Electrification of Cloud and Raindrops

BENJAMIN FRANKLIN's pioneer experiments and success in extracting electricity from thunderclouds in the year 1752 initiated 200 years of speculation as to the magnitude and causes of rain-cloud electrification. Franklin's direct resort to experimentation was, of course, a mark of his genius because at that time the capabilities of the powerful team of scientific experiment and mathematical analysis were but dimly recognized.

Within the last two years our laboratory has made tremendous advances, both experimentally and in an understanding of the basic processes responsible for the observed electrification of cloud droplets and rain. It is our purpose here to review and summarize the principal experimental and analytical results. Space does not permit our exploring in detail the important role that electrified drops and droplets play in the production of lightning, the stability of clouds, the control they exercise in determining rates of precipitation, and their function in removing the fine particle pollution normally present in the atmosphere. One may note only that droplet electrification influences all these matters to an appreciable extent, and their exact relationships may be investigated once a quantitative understanding of the electrification processes of cloud droplets and rain is available. Benjamin

Franklin would have immediately recognized the importance of these matters had he access to the modern facts on this subject.

Reliable observational data concerning the electrification of cloud and raindrops have been scarce and difficult to obtain. Accordingly, our laboratory has been actively working to fill this gap in scientific knowledge and to supplement the new measurements by careful analyses of the basic processes. It appeared a few years ago that little or no progress could be hoped for in the understanding of the basic electromechanics of rain cloud electrification because of the apparent complexity of the problem. However, recent studies provided clues that have reduced the fundamental processes to quite simple terms and these succeed in describing, in a highly satisfactory way, practically all the observed phenomena. One first considers the fundamental electrification processes in the clear atmosphere.

1. ELECTRICITY OF THE CLEAR ATMOSPHERE

It is well known that any electrically charged and highly insulated conductor in the atmosphere systematically loses charge to the surrounding air. This loss shows that the air is a poor electrical conductor, and it is a well-established fact that this conductivity results from the presence and motion of ions produced in the atmosphere by cosmic rays and local radioactivity. Under unusual conditions the normal conductivity may be supplemented by other processes. Measurements show that in clear weather a negative electrical charge approximating 4×10^{-4} esu per cm^2 resides on the surface of the earth and that a current of downwardly moving positive ions is thereby maintained of such magnitude that the surface charge would be largely neutralized in about 500 sec unless systematically replenished. From measurements of this current and the surface electric charges we may determine that the normal conductivity near the surface of the earth approximates 2×10^{-4} esu. Such a conductivity implies that about 700 highly mobile ions are normally maintained in a cubic centimeter of the air by some ionizing agency. Roughly 10 ion pairs per cm^3 sec are continuously produced near the surface by the ionizing radiations. It is important to notice that both a positive and negative ion are always generated simultaneously.

The aforementioned quantities are descriptive of the electrical state at the earth's surface. However, at increasing altitudes the ionic mean free paths increase as does the rate of ion production by cosmic rays. Therefore, the electrical conductivity and ionic densities systematically increase to a considerable altitude and at 5 km a charged body loses most of its charge in 140 sec. The mean ionic density at this level approximates 1100 ions per cm^3, while the rate of ion pair production approximates 12 ion pairs per cm^3 per sec. It may be noticed, therefore, that the ionic population and its rate of generation are considerable at ordinary rain-forming levels. These ions have a profound influence on the electrification of cloud droplets.

2. ELECTRICITY OF CLOUDS

The generation of cloud droplets in the earth's atmosphere produces marked changes in the normal clear air electrical state. The electrical conductivity within a stable cloud is much less than in the clear air, and charges are normally observed to collect on the droplets.

Several measurements have been made on the charges carried by natural cloud droplets. For example, both Wigand [14] and Scrase [12] have measured charges on the cloud elements particularly when the cloud was in the nature of a wet fog. More recently, Gunn [6] explored the matter using airborne equipment to sample independently the free charges carried by cloud droplets and by the associated air. The apparatus shown in Fig. 1, consisting of a centrifuge for separating out the droplets from the environmental air and a device for capturing the remaining ions, was installed in the nose of a B-25 bombing plane. Clouds of many different types were analyzed. It was found that most nonprecipitating clouds were essentially neutral and that the net charge carried by the larger cloud droplets was opposite in sign to that on the associated air and water molecules. When the clouds were slightly unstable, the net charge on the droplets was sometimes observed to be positive and sometimes negative, whereas the measured space charge droplet densities were typically 5×10^{-6} esu per cm^3. Measurements with somewhat similar equipment at ground levels by Webb and Gunn [13] showed a similar type of electrification and suggested that the *net* charge on cloud droplets is always quite small except when the clouds are precipitating.

It became clear from these measurements within natural clouds that the average charge on the droplets was not particularly significant and what was required was a complete analysis of the distribution of charges on cloud droplets. There are serious practical difficulties in obtaining such measurements in the free atmosphere, but we have

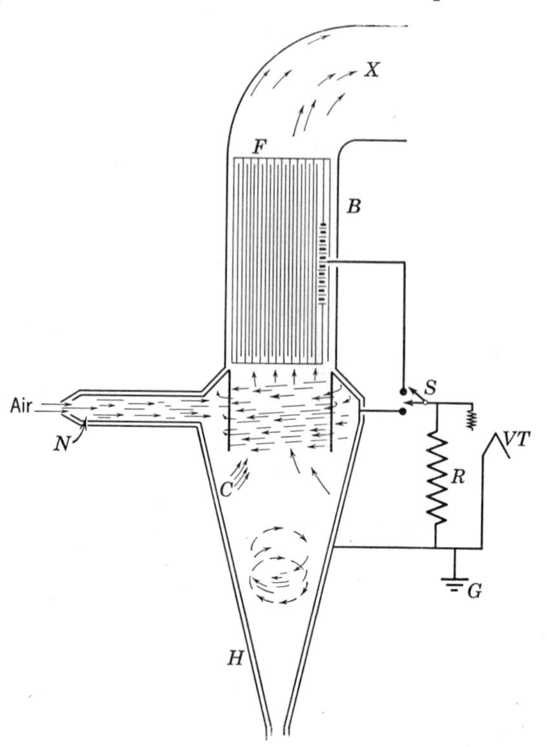

Fig. 1. Schematic diagram of cloud droplet centrifuge *C* and ion filter *F* mounted in the nose of a B-25 airplane and connected to current amplifier *VT*. The entire unit is shielded by conductor *H–B*. Apparatus permits separate determinations of the free charge carried by cloud droplets and by the associated air.

been entirely successful in determining the charges on cloud droplets artificially produced in the laboratory. Clouds may be produced in the Weather Bureau's giant cloud chamber that reproduce the processes of nature and these clouds have enabled us to attack the problem in a perfectly straightforward manner under favorable controlled conditions. This giant cloud chamber is shown in Fig. 2. By letting individual cloud droplets fall through a smaller chamber that is pervaded by a strong horizontal electric field which can be varied cyclically both in magnitude and sign, it is possible to determine from the motions imposed on a falling cloud droplet the magnitude and sign of its free charge together with a reliable estimate of its mass. By producing a typical cloud and letting a few hundred droplets fall successively through such a chamber, it is possible to photograph their

Fig. 2. Giant cloud chamber of the Weather Bureau designed to bring cloud processes under scientific control within the laboratory.

trajectories and determine the distribution of the fractional numbers of droplets carrying any selected free charge. The first distribution curve of cloud droplets obtained using this technique is shown in Fig. 3. It will be seen that somewhat more than half the droplets carry positive charges, whereas the remaining fraction carries negative charges. The slight observed asymmetry results from different values for the positive and negative light ion conductivities inside the cloud.

A long series of similar measurements on droplets of various types has provided a capital clue to the basic mechanisms responsible for the droplet electrification. For example, Fig. 4 shows the measured

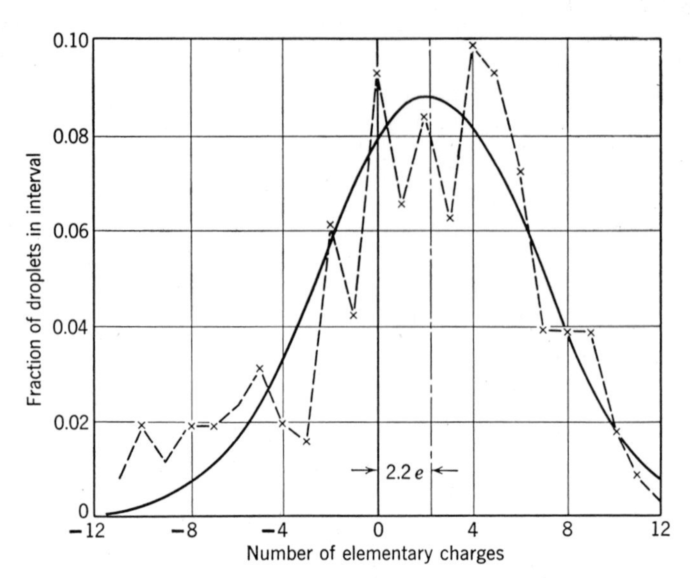

Fig. 3. Distribution of the fractional number of droplets in relation to their charge. Observed data on 250 cloud droplets of mean radius 1.15×10^{-4} cm (dashed curve). Solid curve is calculated from Eq. 2 (p. 12).

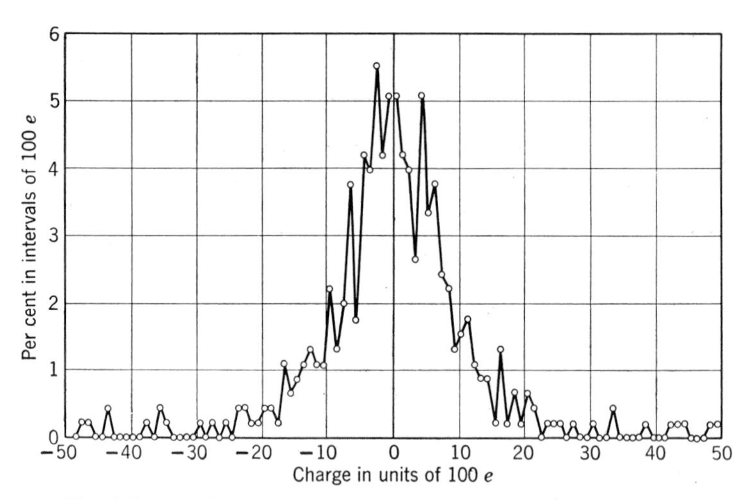

Fig. 4. Cloud formed by sprayed water. Distribution of charges immediately after dispersal. Notice the large number of highly charged particles.

distribution of charges for cloudlike droplets produced by the energetic spraying of ordinary tap water. It may be noticed that large numbers of highly charged particles are produced as well as a rough distribution of weakly charged particles.

Entirely similar effects are measured if the spray cloud is replaced by various dust clouds dispersed into the air by an energetic blast of air. Further measurements were made on clouds produced in the cloud chamber by the condensation of water vapor. It was at once observed that such cloud particles, when initially formed, were essentially neutral and very few droplets had as much as one ion on them. It has been observed that when any of these clouds were exposed to X-rays or other ionization sources they very shortly became electrified and exhibited a distribution as shown in Fig. 5, the curve *B*. Measurements of these general types were made on a number of different clouds produced in a variety of ways, and *it was found that no matter what the state of electrification of the aerosol was initially, upon exposure to copious ionization, a distribution shortly developed that was essentially Gaussian in nature and similar to the theoretical curves given in Figs. 5 (the curve* A) *and 6.* This demonstrates very clearly that a single basic process is responsible for the equilibrium

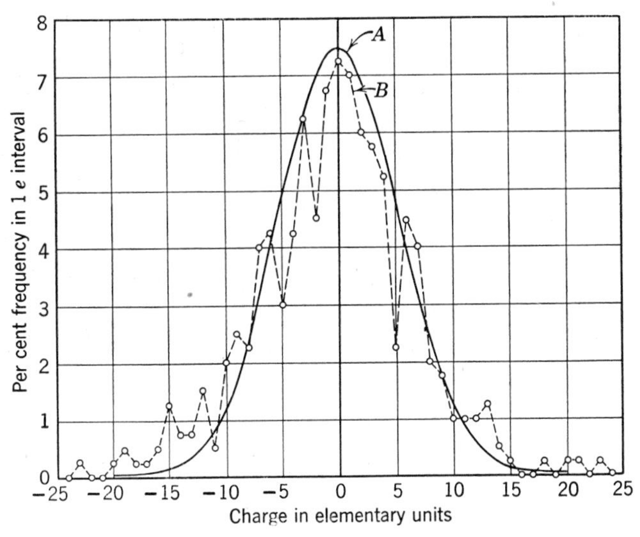

Fig. 5. Water cloud formed by condensation of vapor. This condensed cloud was initially uncharged (radius 1.63×10^{-4} cm). Dashed curve *B* gives distribution after droplets were exposed to copious ionization. Solid curve *A* is calculated distribution according to Eq. 2.

electrification of all aerosols and clouds and that the process is somehow related to the presence of ions in the atmosphere.[15]

A cloud consists of large numbers of droplets, and measurements show that, although each individual cloud droplet may carry from 1 to 20 or 30 elementary charges of a single sign, the distribution of these charges among the droplets is generally such as to make the cloud, as a whole, neutral. That is to say, equal numbers of positively and negatively charged droplets normally exist in a typical cloud. This situation is made manifest by electric field measurements within natural clouds which show that stable, nonprecipitating clouds are essentially neutral and the electric field in their vicinity is small. Unstable associating clouds, on the other hand, usually exhibit a measurable *net* electrification.[6]

3. THE ELECTROMECHANICS OF DROPLET CHARGING

An analysis of the data summarized in the foregoing paragraphs and comparison with detailed theoretical estimates show very clearly that cloud droplets are primarily electrified by the diffusion of atmospheric ions onto the droplets. In general, both positive and negative light ions are present in the atmosphere and are reasonably abundant at cloud-forming levels. Thus, the known thermal jostling of the ions normally results in a transfer of selected ions to the surfaces of the various droplets. The number and rate of charging of the droplets are clearly determined by probability considerations.

Two independent droplet-charging regimes by ionic diffusion are immediately evident,[3,8,9] namely: (*a*) Systematic charging that occurs when the probability of a positive ion striking a droplet is systematically different from the probability of a negative ion striking the same droplet; (*b*) Random charging that normally results when cloud droplets are bombarded by positive and negative ions under conditions such that the probability of the capture of a positive ion is exactly the same as the probability of capture of a negative ion. It should be noticed, however, that as a cloud droplet accumulates a considerable number of ions of a single sign, the probability of capture of ions having the same sign becomes systematically less due to the repulsion. In the same way the accumulated charge systematically attracts and thus increases the probability of capture of ions having an opposite sign. The magnitude of the equilibrium charge is thereby limited.

4. SYSTEMATIC CHARGING

Consider a cloud droplet in the ionized atmosphere. The droplet is surrounded by both positive and negative ions that bombard the droplet as a result of their thermal motions. The probability of a positive ion striking a droplet is proportional to both the number of ions per unit volume and to the speed with which the ion diffuses which is, in turn, measured by the electrical mobility of the ion. Thus, if the product of ion density and mobility is the same for both positive and negative ions, the long time average of the charge on the cloud droplet will be zero. However, if for any reason the density and mobility product for either ion exceeds that of the opposite type, the droplet will acquire and maintain a positive or negative *net* charge. In a recent paper [8] the detailed electromechanics of these charging processes has been worked out and it has been shown that the free systematic charge Q on a freely falling droplet is given by

$$Q = \left[1 + F\left(\frac{aVe}{2\pi kTu}\right)^{1/2}\right]\frac{akT}{e}\ln\left(\frac{n_+u_+}{n_-u_-}\right) \tag{1}$$

where a is the radius of the droplet, k is the Boltzmann constant, T is the absolute temperature, e is the elementary ionic charge, V is the velocity of fall, u the mean ionic mobility in the transition layer, and u_+ and u_- are the respective mobilities. This expression has been tested in the laboratory using a small wind tunnel employing ionized air as a carrier, and the agreement of Eq. 1 with observation has thereby been established.

5. RANDOM CHARGING

The random charging of cloud droplets is important when the probability of capture of a positive ion is exactly the same as the probability of the capture of a negative ion. Therefore, the charge on a particular droplet averaged over a long period of time will be zero. However, statistical fluctuations constantly occur, and at any given instant there is a high probability that any particular droplet will have an accumulation of ions. Because the probabilities of ion capture are the same for both types of ions, it is clear that for every positive droplet carrying a given charge there is likely to be a similar negatively charged droplet in the same vicinity.

6. FUNDAMENTAL AEROSOL DISTRIBUTION

The author's quantitative investigation of the probabilities of ion capture by cloud droplets has shown that a Gaussian-like distribution of charged cloud droplets is shortly established.[9] The fractional number of cloud droplets carrying x elementary charges is given by *the fundamental aerosol-electrification equation*

$$\frac{F_x}{F_t} = \left[\frac{e^2}{2\pi akT}\right]^{1/2} \exp\left[-\frac{\left[x - \frac{akT}{e^2}\ln\left(\frac{n_+u_+}{n_-u_-}\right)\right]^2}{2\frac{akT}{e^2}}\right] \tag{2}$$

where F_t is the total number of droplets per unit volume, F_x is the number per unit volume carrying x elementary charges, a is the droplet radius, e is the elementary charge, k the Boltzmann constant, T the absolute temperature, n_+ is the ionic density for the positive light ions, and u_+ is their mobility. It may be noticed that n_+u_+/n_-u_- is also the ratio of the positive and negative light ion conductivities. Plots of Eq. 2 for droplets of various sizes are shown in Fig. 6. This fundamental aerosol-electrification equation includes both the systematic charging represented by Eq. 1 (when V is small) and the random electrification just mentioned. Many laboratory measurements confirm the essential correctness and reality of Eq. 2.

When the systematic electrification is zero, corresponding to $n_+u_+ = n_-u_-$, this equation degenerates to a symmetrical Gaussian form from which it is easy to show[9] that the average charge on both the positive and negative *fractions* of the cloud droplets is

$$\bar{q}_+ = \bar{q}_- = \left[\frac{\pi akT}{2}\right]^{1/2} \tag{3}$$

By rewriting this expression it is interesting to notice that an equipartition is established between the mean electrical potential energy carried by each droplet and the mean thermal kinetic energy of the bombarding ions.[9] This is an important consequence of our investigations and a somewhat similar equipartition will be mentioned in connection with the electrification of rain.

The mathematical analysis necessary to establish Eq. 2 is quite complex but the basic physics is easily understood. To illuminate this

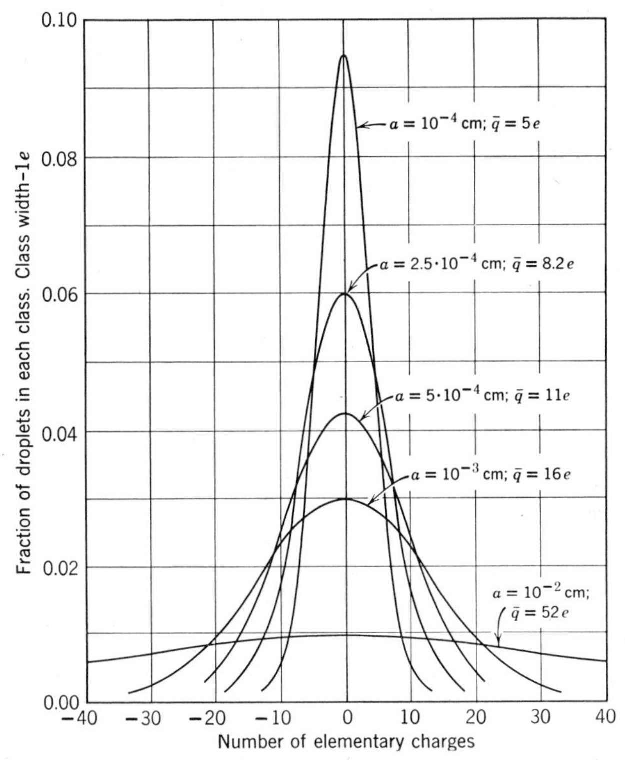

Fig. 6. Calculated distribution as deduced from Eq. 2 of charged cloud droplets having the radii specified on each curve. $\bar{q}$ is the average number of elementary charges, irrespective of sign.

fundamental process one may refer to Fig. 7 wherein the charges communicated to a group of cloud droplets may be quickly estimated on the assumption that the probability of capturing a positive ion is exactly the same as the probability of capturing a negative ion. Starting for convenience, with 1024 droplets it is evident that after each droplet captures 1 ion, half of them will have 1 positive ion and the other half 1 negative ion. If these two groups capture a second ion, it is clear that half of each group will capture a positive ion and half a negative ion. This results in 256 of them having 2 positive ions while 256 will have 2 negative ions and 512 of them will be neutral. Furthermore, if each one of these new groups captures a third ion, it is clear that 128 will have 3 positive ions and 128 will carry 3 negative ions while half of the remainder will have 1 negative ion and half 1 positive ion. By following through such a scheme as shown

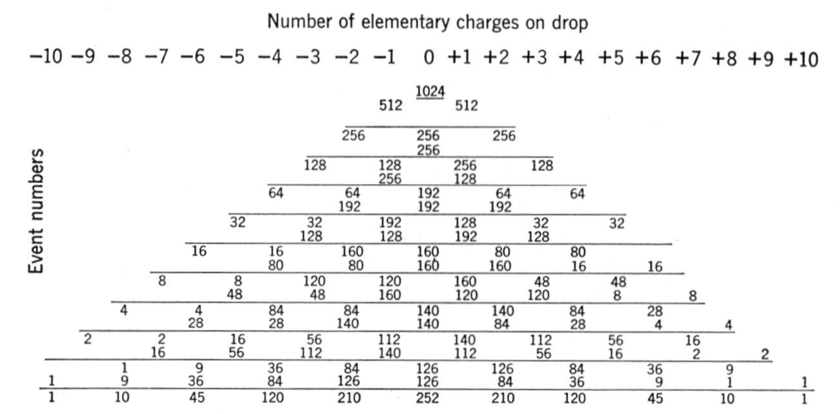

Fig. 7. Schematic representation of the number of cloud droplets carrying indicated numbers of positive and negative elementary charges as a function of the number of collision events. The final distribution on the bottom row corresponds to the binomial point expansion of statistical theory when the probability of capture of positive and negative ions is the same.

in Fig. 7 it may be seen that the distribution after 10 collisions is that shown. This distribution corresponds to the binomial point distribution of statistical theory. It is a well-known fact that this distribution approaches the Gaussian normal distribution as the number of capture events increases to large values. Thus, one may understand the fundamental electromechanics underlying the derivation of Eq. 2.

The reader may find it of interest to construct a diagram like Fig. 7, wherein the probabilities of capture for the positive and negative ions are somewhat different, and satisfy himself that the resulting distribution will be skewed toward a greater abundance of droplets having a sign the same as the predominant electrical conductivity. Such an analysis will show that the resulting distribution closely approximates that given by Eq. 2.

7. ELECTRICITY OF RAIN

The author has believed for many years that the compilation of data on the free electrical charges brought down by rain would illuminate the fundamental problems of cloud and raindrop electrification. Considerable effort has, therefore, been devoted to the invention and development of new apparatus and new techniques. Fig. 8 shows an apparatus that was built to collect data on the size and charge carried by rain. The apparatus is so designed that an untouched drop

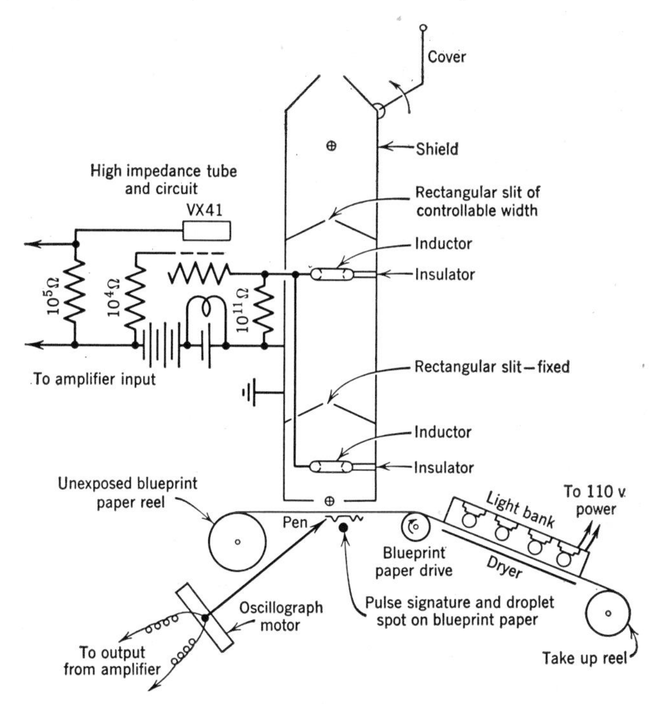

Fig. 8. Induction apparatus for recording the sign and magnitude of the free charge on freely falling raindrops together with their velocity of fall and mass.

from the sky falls through a collimating tube and two highly insulated inductor rings. When the charged raindrop passes the first ring, it produces a pulse in an amplifier that is amplified and passed to the oscillograph. After falling another meter, the same drop passes through another inductor and puts a second pulse on the oscillograph paper. The oscillograph paper is pulled through the oscillograph at a uniform speed and consists of ordinary blueprint paper upon which the raindrop ultimately falls to make a spot. This spot is developed and made permanent by the indicated light bank and the paper is then wound on a reel. With this apparatus the sign and magnitude of the produced pulses determine the charge on the droplet. The time it takes for the droplet to fall between inductors separated by a meter determines the velocity of fall of the droplet, and the size of the spot permanently recorded on the blueprint paper determines the droplet mass. The complete history of the drop is therefore recorded.[5] This apparatus, and modifications suitable for use on aircraft, has been of great value in our raindrop electrification studies.

The first investigation of the charges carried by individual raindrops was reported by Gschwend [2] who also measured the size of the drops. His early measurements of a relatively few drops have been generally verified by later observers, and these all show that the free charge per droplet at the ground is usually of the order of 10^{-3} esu, except for electrical storm rain when it sometimes approximates 10^{-2} esu. An outstanding characteristic noted by Gschwend and other observers is that a mixture of both positive and negative droplets normally is observed below all rain clouds and that after one or two drops of a single sign are captured there is a high probability that the next drop will be of opposite sign. It is fair to remark that this fact has puzzled geophysicists for a good many years but the electromechanics underlying these observations will be made clear presently.

A long series of measurements of drop charge by Chalmers and Pasquill [1] gave average values as summarized in Fig. 9. More recently, using the apparatus described in Fig. 8, Gunn [5] and Gunn and Devin [7] have made further measurements which generally confirm the earlier reported values. The available data on raindrop charges are summarized in Table 1.

Since the atmosphere through which the rain is falling is slightly conducting, electrified raindrops discharge as they fall. Calculation shows that a raindrop falling 2 or 3 km in clear air will discharge a large fraction of its initial electrification. Therefore, charge measurements at ground level are not very significant. In order to determine the magnitude of the free charges on rain at the rain-forming level, it was necessary to fly into such regions with experimental aircraft and specially developed electrical equipment. The Army-Navy Precipitation Static Project that the writer directed at Minneapolis, Minnesota, during the war, provided an opportunity to make such measurements. A modification of the induction method apparatus shown in Fig. 8 was developed to measure the charges on raindrops. Under the right wing of a B-17 airplane was mounted a highly insulated inductor ring that was protected from rain and cloud droplets by a truncated sheet metal cone attached to the airplane. The small circular opening of this cone that faced into the oncoming precipitation permitted raindrops to pass through both it and an internally mounted and highly insulated inductor ring. A large fraction of the raindrops would traverse the metal cone inductor ring without touching either. A few drops would strike the leading edge, splatter, and produce a signature on an oscillograph that could be easily separated out from the signature of raindrops that traversed the inductor ring without

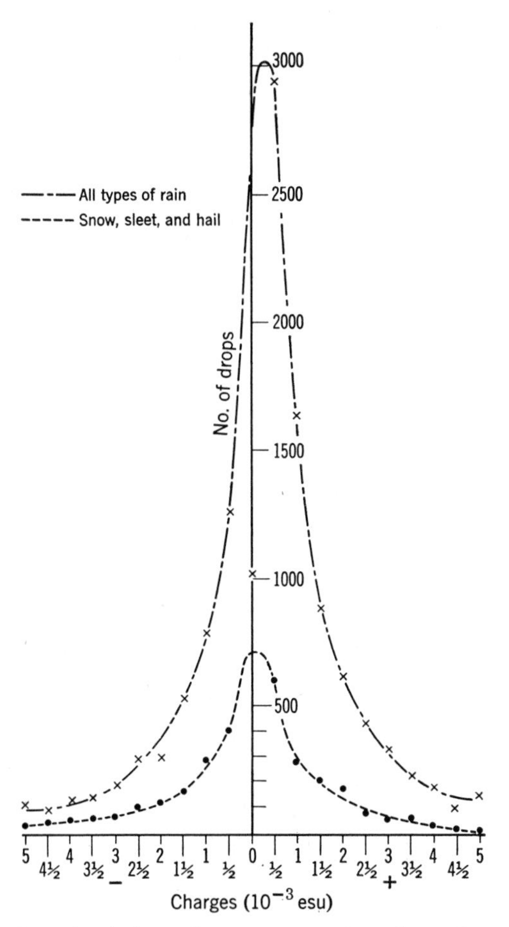

Fig. 9. Distribution of raindrop charges measured at the surface of the earth in England by J. A. Chalmers and F. Pasquill.

touching it. Charged raindrops passing through this ring induce an electrical pulse on the ring that can be amplified and passed to an oscillograph that permits the measurements of the sign and charge carried by the original raindrop.

In a mild cold front, near Minneapolis, coherent measurements were made on the charges carried by raindrops in rain-forming regions and at a number of different levels. These data have been published,[4] are summarized in Table 1, and their distribution plotted as a function of the drop charge in Fig. 10. It is important to notice that the charges measured at the rain-forming levels are some 10 to 30 times

TABLE 1. AVERAGE FREE ELECTRICAL CHARGE ON INDIVIDUAL DROPLETS

(ESU × 10^3)

Observer	Altitude (ft)	Charge	Quiet rain	Shower rain	Electrical storm rain	Quiet snowfall	Squall snowfall
Gschwend (1921)	surface	+	0.24	1.75	8.11	0.09	5.64
		−	0.53	5.43	5.88	0.06	4.78
Banerji and Lele (1932)	surface	+		6.4	6.9		
		−		6.7	7.3		
Chalmers and Pasquill (1938)	surface	+	2.2	1.3	3.7 *		10.5
		−	3.0	2.3	9.2 *		5.7
Gunn (1947)	4,000	+		†			
		−		24			
	12,000	+		41			
		−		100			
	20,000	+		63			
		−		†			
Gunn (1949)	surface	+			15	0.67	
		−			19	1.0	
Gunn (1950)	5,000	+			81		
		−			63		
	10,000	+			148		
		−			112		
	15,000	+			123		
		−			76		
	20,000	+			52		
		−			62		
Gunn and Devin (1953)	surface	+			22		
		−			31		

* Actual lightning activity doubtful.

† No droplets of this sign were observed at the indicated level.

greater than those typically measured at the ground and that a mixture of positive and negative drops was usually present. Raindrop charge analyses were carried out in both shower rain in which no appreciable vertical convection or electrical activity was noticed and also in active thunderstorm electrical conditions.

An outstanding characteristic of the charges measured at active rain-forming levels is that the charges on many of the drops are so great that the electric field at their surface is an appreciable fraction of the dielectric strength of air. Thus, many of the drops are as highly electrified as they can possibly be. Moreover, it is found that in such clouds, as well as at the earth, a roughly Gaussian distribution of droplet charges is established. The solid line of Fig. 10 shows the distribution of drop charges as measured at the rain-forming levels in a mild cold front in Minnesota on July 27, 1945.[4] Thus, within a cloud there is some process which systematically places exceedingly large charges on some droplets and exceedingly large charges of opposite sign on other drops. The basic physics that could produce such a distribution has been obscured for a long time. However, a well-fitting key to the problem is now available, and one turns to a consideration of one of the fundamental processes whereby rain is electrified.

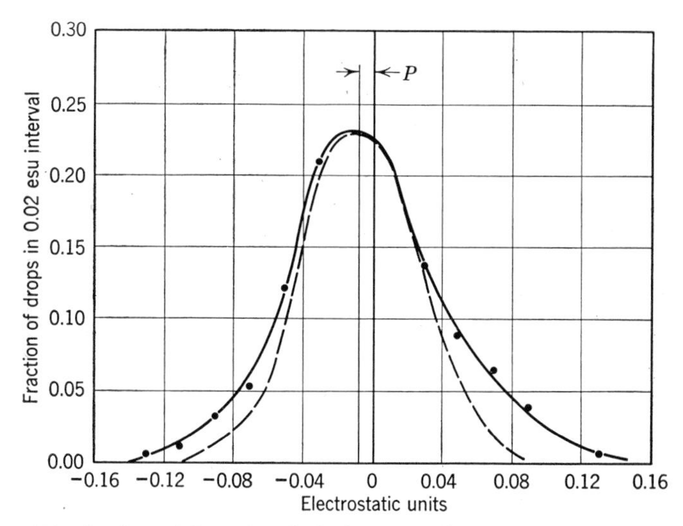

Fig. 10. Distribution of free electrical charge on "moderate rain" within a mild cold front in Minnesota on July 27, 1945. Measurements made at or near rain-forming levels by use of aircraft.

According to the foregoing discussion a typical cloud consists of large numbers of droplets in which about half of the droplets carry a positive charge that is typically 11 elementary units, while the other half typically carries 11 negative units. Normally 4 per cent or less of the droplets are uncharged. Suppose that a small raindrop falls down through such a cloud and grows by association with the cloud droplets. The falling raindrop is accordingly bombarded in a purely random manner by cloud droplets which it successively intercepts. If there are equal numbers of positively and negatively charged cloud droplets, the probability of the drop colliding with a positive droplet is the same as the probability of its colliding with a negative droplet. The statistical distribution of charge may, therefore, be worked out in a manner precisely like that used to determine the charge on the cloud droplet when it was bombarded by ions. The fundamental difference between the electrification of the cloud droplet and the raindrop is that the cloud droplet encounters ions by their thermal agitation, whereas the raindrop is bombarded as a result of the relative gravitational motions of the large raindrop and the small electrified cloud droplets. Two raindrop electrifying regimes by droplet association are evident. The initial or nonequilibrium regime describes the early stages of electrification and is descriptive as long as the probability of collision of a droplet with a charged drop is constant. This regime is gradually converted into the equilibrium regime which is established whenever the accumulated raindrop charges become large enough to control the probability of collision.

Whenever the number of collisions between the raindrop and the cloud droplets is limited and the probability of collision is constant, the average charges *for the distributions* illustrated in Fig. 7 may be estimated by the binomial point equation of statistical theory. The mean *nonequilibrium* charge accumulated on raindrops $\hat{Q}$, *averaged without regard to sign*, is nearly

$$\hat{Q} = \left[\frac{2K}{\pi}\right]^{1/2} \bar{q} \qquad (4)$$

where K is the number of collisions that establish the charge on the raindrop and $\bar{q}$ is the mean charge on the parent droplets irrespective of their sign. Since K may be large, it is clear that very large drop charges may sometimes accumulate.

In the special case where there are more cloud droplets carrying one kind of charge than the other, and their numbers per unit volume are C_+ and C_-, one may show that the mean *systematic* charge $\bar{Q}$ accumulated on the raindrops, *averaged with respect to sign* is nearly

$$\bar{Q} = \frac{K\bar{q}}{2} \ln \left(\frac{C_+}{C_-}\right) \tag{5}$$

The above nonequilibrium expressions for raindrop charges are applicable only when the number of collisions is known and the accumulated charges are inadequate to modify appreciably the probability of droplet collision.

Whenever there are sufficiently large numbers of collisions with the cloud droplets so that the charges accumulated by the raindrop become large enough to repel or attract the charged cloud droplets and thus modify their probability of collision, it becomes necessary to work out the *equilibrium* distribution. A mathematical analysis of this complex problem shows [10, 11] that the distribution is much like that given by Eq. 2 or

$$\frac{D_{\Omega q}}{D_t} = \frac{\bar{q}}{\left[\dfrac{\pi(r_1 + r_2) U_R^2}{2(1/m_1 + 1/m_2)}\right]^{1/2}} \exp \left[\frac{-\left[\Omega\bar{q} - \dfrac{(r_1 + r_2) U_R^2 \ln (C_+/C_-)}{4\bar{q}(1/m_1 + 1/m_2)}\right]^2}{\dfrac{(r_1 + r_2) U_R^2}{2(1/m_1 + 1/m_2)}}\right] \tag{6}$$

where D_t is the total number of raindrops per unit volume, $D_{\Omega q}$ is the number per unit volume carrying a charge $\Omega\bar{q}$, $\bar{q}$ is the mean cloud droplet charge irrespective of sign, Ω is an integral number, m_1 and m_2, and r_1 and r_2 are the masses and radii of the cloud and raindrop, respectively, and U_R is the relative velocity of the two types of drop. It may be noticed that this expression is similar in form to Eq. 2 and is analogous to it in many ways. Using this expression one may calculate the mean charge on both the positive and negative *fractions* of the falling rain and show thereby that this charge is given by

$$\overline{\Omega\bar{q}} = \overline{Q}_+ = \overline{Q}_- = \left[\frac{\pi(r_1 + r_2) U_R^2}{8\left(\dfrac{1}{m_1} + \dfrac{1}{m_2}\right)}\right]^{1/2} \tag{7}$$

As in the case of Eq. 3, this expression may be rewritten to show that an approximate equipartition is established in which the electrical potential energy carried by the average falling raindrop is equal to the energy of bombardment of the falling raindrop by the smaller cloud particles. Accordingly, the mean charge, irrespective of sign, is determined principally by the mass of the smaller cloud droplets and the relative velocity of the raindrops and cloud elements. It may be noticed that the *equilibrium* charge on the raindrops does *not* depend upon the charges carried by the cloud droplets and this quantity influences only the fraction of droplets carrying a given charge.[11]

It should be clear from a consideration of the above basic mechanisms that the electrification of droplets at the rain-forming level must be much greater than at the earth and that nearly equal numbers of positive and negative droplets are likely to be captured by any measuring equipment. This is exactly what is observed. By assuming that the rain was "medium rain" and that droplets of 0.05 cm radius fell through a cloud of mist-sized droplets of radius 0.005 cm, one may calculate a curve according to Eq. 6, and this is plotted as the dashed curve in Fig. 10. A comparison of the observed and calculated curve shows that the agreement is perhaps better than one has a right to expect, since the basic assumption that there were drops of only two sizes certainly is not exactly true in nature.

In Fig. 11 the results of the foregoing analysis have been summarized. This figure gives the mean equilibrium droplet charge, irrespective of sign, that may be expected on falling rain at the rain-formation level. The raindrops of radii corresponding to the abscissa are assumed to fall through various uniform clouds having the droplet radii specified on each curve. The resulting equilibrium electrical charge is then given by the ordinate. An examination will show that the largest raindrops are so highly charged that they would discharge by corona if further charge were added to them. This surprising state of affairs was first noticed by the author in 1947 [4] and is adequately explained by the previously considered electromechanics of drop electrification.

8. CONCLUSIONS

The rather simple concept of falling raindrops and cloud droplets being statistically bombarded by charged cloud particles and ions has been exceedingly fruitful in describing the known complex electrical characteristics of rain and cloud droplets. The distributions closely

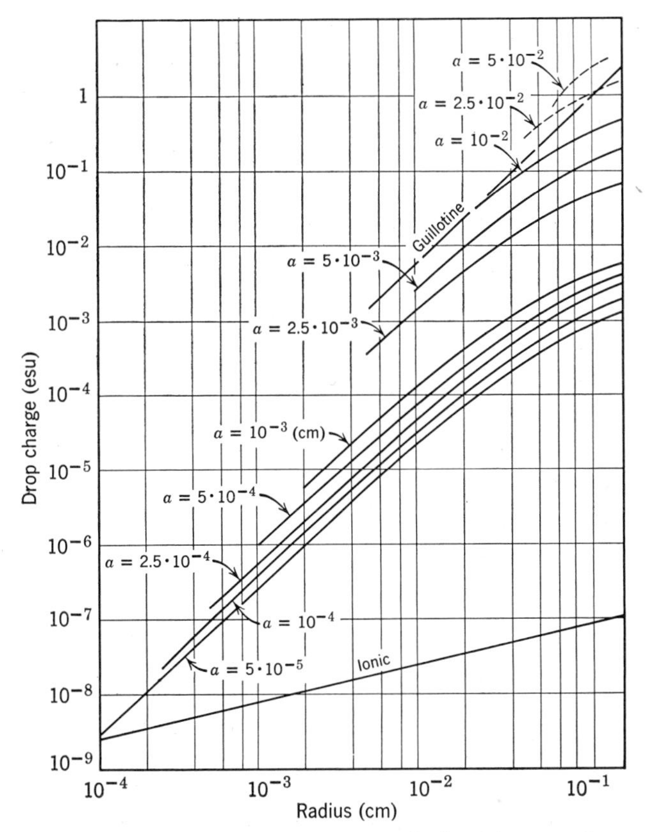

Fig. 11. *Equilibrium* free charge on raindrops calculated in accordance with Eq. 7. Each curve corresponds to a cloud of smaller droplets (having mean indicated radius a) through which larger drops, of radius given by abscissa, fall. Lower straight line is equilibrium charge placed on droplets by thermal diffusion of the atmospheric ions as estimated from Eq. 3. Drop charge corresponding to a surface electric field of 15,000 v/cm is labeled "Guillotine." These values apply only at the rain-formation level and must be corrected for discharge to represent the values after the drops fall to the surface.

approximate those normally observed at rain- and cloud-forming levels and, when corrected for electrical discharge as they fall in the free atmosphere, well describe the observed electrifications at the earth's surface. A series of investigations now being prepared for publication shows that the calculated electrification is adequate, when combined with new influence processes, to describe most of what is now known about thunderstorm electricity and related problems. The concepts are capable of extension to the description of volcanic lightning and, indeed, to electrical storms on other planets or on the stars. Because of the universal character of the electrification processes, the effects are likely to be present in a large number of cosmic and terrestrial phenomena.

REFERENCES

1. Chalmers, J. A., and F. Pasquill, The electric charges on single raindrops and snowflakes, *Proc. Phys. Soc. (London)*, **50,** 1–15 (1938).
2. Gschwend, P., Beobachtungen über die elektrischen landungen einzelner regentropfen und schneeflocken, *Jahrb. Radioakt. u. Elektronik,* **17,** 62–79 (1920).
3. Gunn, R., The electricity of rain and thunderstorm, *Terrestrial Magnetism and Atm. Elec.,* **40,** 79–106 (1935).
4. Gunn, R., The electrical charge on precipitation at various altitudes and its relation to thunderstorms, *Phys. Rev.,* **71,** 181–186 (1947).
5. Gunn, R., The free electrical charge on thunderstorm rain and its relation to droplet size, *J. Geophys. Research,* **54,** 57–63 (1949).
6. Gunn, R., The electrification of cloud droplets in non-precipitating cumuli, *J. Meteorol.,* **9,** 397–402 (1952).
7. Gunn, R., and Charles Devin, Jr., Raindrop charge and electric field in active thunderstorms, *J. Meteorol.,* **10,** 279–284 (1953).
8. Gunn, R., Diffusion charging of atmospheric droplets by ions and the resulting combination coefficients, *J. Meteorol.,* **11,** 339–347 (1954).
9. Gunn, R., The statistical electrification of aerosols by ionic diffusion, *J. Colloid Sci.,* **10,** 107–119 (1955).
10. Gunn, R., Droplet electrification processes and coagulation in stable and unstable clouds, *J. Meteorol.,* **12,** 511–518 (1955).
11. Gunn, R., Raindrop electrification by the association of randomly charged cloud droplets, *J. Meteorol.,* **12,** 562–568 (1955).
12. Scrase, F. J., Electricity on rain, *Geophys. Mem.,* **9** (No. 75), 1–20 (1938).
13. Webb, W. L., and R. Gunn, The net electrification of natural cloud droplets at the earth's surface, *J. Meteorol.,* **12,** 211–214 (1955).
14. Wigand, A., Measuring the electric charges of natural mists, *Z. Physik,* **27,** 803–808 (1926).
15. Woessner, R. H., and R. Gunn, Measurements related to the fundamental processes of aerosol electrification, *J. Colloid Sci.,* **11,** 69–76 (1956).

JOACHIM P. KUETTNER
Geophysical Research Directorate
Air Force Research Center
Cambridge, Mass.

II

The Formation of Electric Charges in Thunderstorms

About 200 years ago Benjamin Franklin wrote a letter to his English friend, Peter Collinson, and described the following experiment (Cohen, 1941):

> In *September* 1752, I erected an iron rod to draw the lightning down into my house, in order to make some experiments on it, with two bells to give notice when the rod should be electrify'd. . . . I found the bells rang sometimes when there was no lightning or thunder, but only a dark cloud over the rod. . . . In the winter following I conceived an experiment, to try whether the clouds were electrified *positively* or *negatively*. . . . At last, on the 12th of April, 1753, there being a smart gust of some continuance, I charged one phial pretty well with lightning, . . . and was convinced that one bottle was electrised *negatively*. . . . I repeated this experiment several times during the gust, and in eight succeeding gusts, always with the same success . . . I concluded that the clouds are always electrised *negatively* . . .

This appears to be the first time that the charge of thunderclouds has been measured. At a time when even the terms, "positive" and "negative electricity" and "charge" and "discharge" had to be introduced by Franklin, he not only proved the identity of lightning and the electric spark, but he discovered already that a thunderstorm cloud appears negatively charged with respect to the ground beneath it. He attacked the problem in a way which is still in use in this century, namely by point discharge currents. One hundred and fifty years passed before his inquisitive research was systematically taken up again.

In a similar project Vincent Schaefer reported in 1947 the observation that in winterly snowstorms the electric field changes its sign periodically over several hours. He called this the "cross current." It is interesting that a similar observation by Franklin, almost 200 years earlier, had cautioned him already to conclude that the electric charge of clouds is only prevailingly negative. In his letter to Collinson of April 1754 he writes (Cohen, 1941):

> Once this winter the bells rang a long time, during a fall of snow, though no thunder was heard, or lightning seen. . . . and last *Monday* afternoon, the wind blowing hard at S. E. and veering round to N. E. with many thick driving clouds, there were five or six successive changes from negative to positive, and from positive to negative, the bells stopping a minute or two between every change.

Earlier he had reported one case of a positive cloud and stated:

> [This] destroys my first too general conclusion, and reduces me to this: *That the clouds of a thunder-gust are most commonly in a negative state of electricity, but sometimes in a positive state.*

This observation has since been confirmed.

It was in 1929, 175 years later, that a man of comparable scientific genius changed the ideas on thunderstorm electricity fundamentally. In his classical Franklin address, the English physicist Wilson, whose thoughts have governed the thinking on this subject for another quarter of a century, said:

> Franklin himself concluded that the surface of the earth below a thundercloud was generally positively charged. . . . And later observations agree with Franklin. As we should now express it, the potential gradients below thunderclouds . . . are much more often negative than positive . . . The view I wish to put before you is a very different one. According to it a thundercloud is essentially bipolar, the positive charge tending to be above the negative. The preponderance of negative potential gradients below shower clouds and thunderclouds is on this view due primarily to the negative charge of the cloud being nearer the ground than the positive.

Wilson was led to this idea by the observation that the sign of the electric field reversed at a certain horizontal distance from the thundercloud. However, he did not stop with this hypothesis. He proposed a mechanism by which this charge dipole is created in the following way. We may again let Wilson speak:

There would probably be general agreement that the source of the electromotive force of a thundercloud is to be sought in the vertical separation under gravity of carriers of positive and negative electricity. Let us suppose that the small particles in a cloud, which only fall slowly relatively to the air, are positively charged, while the larger drops which fall with considerable velocity through the air are negatively charged. We may leave for the present the question how the carriers acquire their charges.

Such a cloud, originally neutral, will at once begin to acquire a positive charge at the top and a negative charge at the bottom through the relative vertical motion of the two classes of carriers. The two equal and opposite charges thus accumulating at the top and bottom may be separated by a great thickness of neutral cloud. The accumulation of a positive charge above and a negative charge below results in the development of an electric field within the cloud.

It should be noticed that Wilson already assumes a decisive role of precipitation in this charging process in accordance with the general meteorological experience. The fact that the precipitation is actually not rain, but overwhelmingly solid in the main portion of the thundercloud, as we know today, does not affect the validity of this part of Wilson's argument. The average temperature of the lower negative cloud charge is generally found to be near $-10°$ C, that of the higher positive charge near $-30°$ C with a wide range of scattering.

In the many years that Wilson observed the electric field, he found that it tends to reach a certain equilibrium value. Lightning then appears more as a disturbance than as a regulating agent, the field recovering to its equilibrium state after each discharge. The exponential time factor varies but is mostly of the order of 5 seconds.

Wilson attempted to build this equilibrium mechanism into his model thus:

The rate of accumulation of the upper and lower free charges and the resulting rate of increase of the field will become less as these charges increase for two reasons. The field within the cloud opposes the falling of the negative drops and the carrying up of the positive cloud particles; and again the dissipation of the upper and lower charges by ionization currents or otherwise increases as the charges increase. There may come a stage where a balance is reached and no further increase in the field results; or on the other hand the field may reach the sparking limit before such a balance can be attained, and we have then a lightning flash.

If such a discharge destroys the vertical field within the cloud while still leaving the positively and negatively charged carriers, the destruction of the field will be followed by its regeneration through the separation of the carriers by gravity; initially at the maximum rate, since no field is opposing the fall of the negatively charged drops relative

> to the positive carriers and there is no loss of charge by ionization currents or otherwise until there are charges to dissipate. The rate of increase of the field will again continually diminish as the upper and lower charges accumulate. If these processes continue to be repeated we must suppose that new carriers are supplied to take the place of those which have been removed.

It seems that, in spite of the strong progress in atmospheric electricity, this particular aspect of Wilson's model has not found the necessary attention. Wilson's charge segregation model and his equilibrium between electric and gravitational forces have been generally accepted, although they are in contradiction to a number of observational facts. This is the more surprising as this mechanism presents a vital link between the much-studied phenomena of charge generation on the one hand and charge distribution and thunderstorm fields on the other hand. As a consequence, conclusions as to the "proper" polarity of charging effects have been built on relatively weak ground.

Wilson himself saw one possible difficulty in the precipitation charge measured at the ground and he took care of this problem in his Franklin address. Obviously the prevailing polarity of the thundercloud (positive–up, negative–down) would lead one to expect that rain carries a negative charge to the ground. It was already known to Wilson that the opposite is generally true. His explanation for this contradiction is:

> The prevailing positive charge on rain is on this view not the cause but the result of the negative potential gradient; the rain intercepts and returns to the earth a portion of the charge carried by the stream of positive ions which are being driven up by the negative potential gradient.

This was actually the starting point of Wilson's well-known ion capture theory which he proposed also as the chief mechanism of charge generation *inside* the thundercloud and which will not be discussed at this point.

1. THE SEGREGATION OF CHARGE

We shall now look a little closer into the charge segregation process and Wilson's arguments.

Let me first describe an experiment especially designed to test Wilson's explanation of the rain charge. If it is true that precipitation particles representing the lower negative cloud charge always lose and even reverse their charge on the long fall from the cloud to the ground, one should expect to encounter their original charge inside the thundercloud close to the lower main charge. However, during experiments on a 10,000-ft mountain in the Alps (Kuettner, 1950) the precipitation leaving the negative cloud charge was found to be positive even at freezing temperatures well inside the thundercloud. While these measurements raised some doubts about Wilson's explanation, there was still the possibility that point discharges from the mountain may have accomplished a quick reversal of the particle charge. The experiment was not designed very well and has not been confirmed by later measurements (Israel et al., 1955).

There is another more serious difficulty with Wilson's equilibrium between electric and gravitational forces. It is known today that the average electric fields in thunderstorms are about one order of magnitude smaller than Wilson thought. There is some argument about whether the average field is closer to 100 volts per cm or to 1000 volts per cm. Dr. Ross Gunn, who has considerable flight experience in thunderclouds, once estimated that the *maximum* field encountered in a thunderstorm averages about 1500 volts per cm and, at one time, he measured close to double this value. We are probably not far from the truth in assuming that the *mean* electric fields inside active thunderclouds are about 600 volts per cm.

This dilemma now develops: If the electric equilibrium is created in the way suggested by Wilson, the gravitational force gm acting on a particle of mass m must equal the electric force $\bar{q}F$ of the electric field F acting on the particle mean charge $\bar{q}$. The specific charge $\bar{q}/m$ then should be of the order of g/F or 500 esu per gram water. It can easily be shown that raindrops of more than 0.3 mm diameter, or spherical snow pellets of more than 1 mm diameter, cannot carry such charge without breaking into corona discharge. It is, however, well established that the common size of precipitation particles in thunderstorms is well above these limits. Also particle charges measured in thunderstorms do not nearly average out at such high net values.

This fact seems to eliminate Wilson's interpretation of the lightning recovery curve whose very small time factor was readily explained by his gravitational equilibrium process. The other alternative, the action of ionic currents, requires very high values of air conductivity—at least 10 times the fair weather conductivity at 20,000-ft altitude in order to identify the recovery time with the ionic relaxation time. While a smaller-than-normal air conductivity is generally expected inside clouds, radar motion pictures seem to give a hint that the lightning discharge itself may enhance the ionic densities over vast cloud areas for short periods. (See Fig. 9.)

There are more difficulties which Wilson's thunderstorm model has to overcome; for example, the positive space charge at the freezing level (the so-called Simpson's charge), the action of up- and downdrafts as found in the Thunderstorm Project, and the fact that the descending precipitation is not falling as a package but is growing and being replaced from higher levels in the cloud.

These considerations make it imperative to study the mechanism of charge segregation more closely. As will be shown now, this mechanism is quite complex and leads to surprising results. The common conclusion that the lower main charge is carried by precipitation and that a correct thunderstorm theory must lead to a negative precipitation charge is not borne out by this study. It is not essential to this investigation in what way the charges are generated, although the most probable mechanism will be taken as an example.

The basic concept of the charge segregation process is the following: let us consider one liter of cloud air in the active center of a thunderstorm. There will be between one hundred thousand and one million super-cooled cloud droplets in this volume, sinking slowly with respect to air and having a spacing of roughly 100 times their radius. About once every second our cloud volume is penetrated by a solid precipitation particle whose diameter is a multiple of the droplet spacing. In the $\frac{1}{50}$ of a second that it takes the ice particle to pass through, it collides with about one thousand super-cooled droplets, the majority of which freeze on contact and are swept out. The precipitation particles growing in this fashion are called "graupel" or snow pellets and are the prevailing type of precipitation in the thunderstorm center. In a later stage of growth they may become hailstones. However, this is by no means necessary. It is generally thought that the charge generation is directly or indirectly connected with this growth and collision process.

It will be clear that the cloud volume must lose the amount of charge withdrawn by the precipitation particles falling through it. If the cloud air is initially neutral, with the onset of precipitation a space charge will appear which is solely represented by the charge of the precipitation particles arriving at this level, no matter how their charge was created. However, the continuous loss of cloud charge to consecutive, growing precipitation particles must result in the build-up of an opposite charge in the cloud air, and it is only a question of time until the cloud space charge reaches the same magnitude as the precipitation space charge at this level. At a certain moment, then, the two charges in the cloud, that on the precipitation and that on the cloud air, will mask each other completely, and the net space charge will vanish temporarily.

The time from the arrival of the first precipitation particles until the total space charge goes to zero will be called the "compensation period." The order of magnitude of this period is only about one minute, as will be shown later. When this compensation period is exceeded, precipitation particles will continue to withdraw charge, and the cloud charge will continue to grow, while the mean precipitation space charge remains essentially constant at a given cloud level, because of the similar life history of consecutive precipitation particles which are passing through. As a consequence, the cloud charge becomes dominant, and the net space charge and the electric field gradient in the areas of active charging will be determined eventually by the sign of the cloud charge, not by the sign of the precipitation charge. However, it cannot be expected that the cloud charge grows indefinitely. The conduction currents set up by the accompanying electric field gradients necessitate the approach of a steady state, where the precipitation and the cloud charges partially mask each other in a certain ratio, but with the cloud space charge always predominating. In this way an equilibrium field forms.

This mechanism works in the described way only in the absence of updrafts. If strong updrafts prevail, the process is modified insofar as the space charge of precipitation in the rising parcel of cloud air is not quasi-constant, but increases systematically with height for a twofold reason: First, more particles are activated and charged and, hence, encountered by the ascending cloud parcel as it reaches higher levels. Second, the size and charge of individual precipitation particles are increasing while the particles are rising in the upcurrent, provided the rate of ascent of the air is greater than the rate of descent of the precipitation. As a consequence, the precipitation space charge in the rising air parcel may increase more quickly than the cloud charge and no compensation time is reached. Thus, the sign of the total space charge in the upcurrent will be that of the precipitation, although it is partly masked by the compensating cloud charge.

Since, in the mature state of a thunderstorm cell updrafts prevail only in the upper portions of the cloud (Byers and Braham, 1949), Fig. 1, it should be expected that the polarity of the upper charge will be essentially that of the precipitation, while the polarity of the lower cloud will be that of the cloud air. The development of a

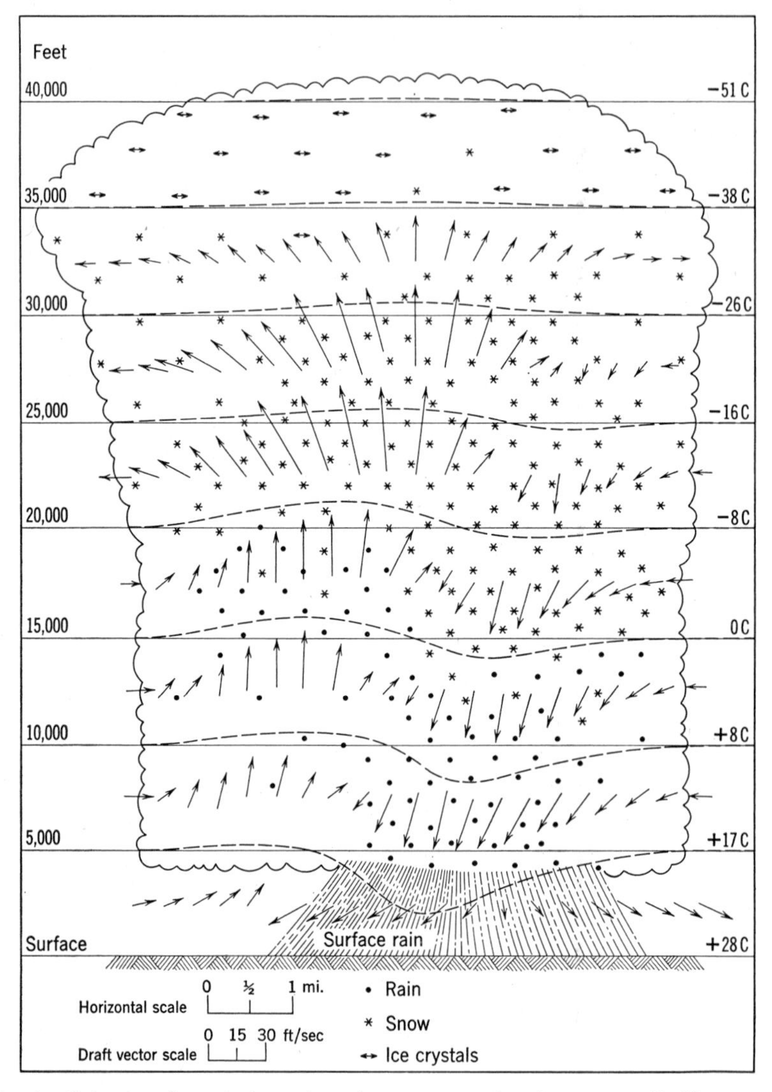

Fig. 1. Scheme of vertical motions in a mature thunderstorm cell (Byers and Braham, 1949).

dipole thus can be connected to the development of the drafts. The charge distribution will be in agreement with that observed in the average thunderstorm (positive–up, negative–down), if the prevailing charge of the ice precipitation in nature is positive in sign. This tentative conclusion is at variance with the generally held view.

However, there is no escape from facing the problem presented here, which makes the very challenging results from experiments on ice electrification take on a new interest. It is recalled that some findings, like those of Findeisen, 1940, have not been accepted because they were thought to give "improper polarity." Since it is most difficult to duplicate the natural conditions accurately in the laboratory, more extensive investigations in natural super-cooled clouds will be of increasing importance, such as those which Findeisen, 1940, and Lueder, 1951, attempted. The Mount Washington Observatory therefore set up an experimental project in 1954 to clarify questions along this line under a variety of conditions and over a long period of investigations. Preliminary results show both polarities occurring on natural rime with the positive sign preferred. The charging rates are higher than expected, about 10^{-11} to 10^{-12} amp per cm^2 of exposed surface.

At certain levels, mostly determined by temperature, the charging process will have to stop. Outstanding among these "stoplevels" will be the freezing level and the –40° C (self-nucleation) level, where all liquid water freezes.

The effect of a "stoplevel" thus is to "unmask" a considerable part of the hidden precipitation charge. As a consequence, a change in sign of the electric field gradient is produced at the stoplevel. "Unmasking" also occurs if the precipitation intensity fluctuates, especially if the shower activity is temporarily interrupted, fully exposing the cloud charge. This may account for high local fields and the initiation of lightning discharges.

In the neighborhood of the freezing level, charging by ice formation discontinues. If no alternate charging effect takes over, the precipitation particles will lose charge by air conductivity below the melting level during their further fall. In contrast to the situation above the freezing level, the precipitation charge now dominates the cloud charge of opposite polarity. This may be the explanation of Simpson's positive charge in the cloud base which generally coincides with the precipitation center. So much for the qualitative description of the charge segregation and masking process. A rigorous quantitative derivation has to follow the variations of space charge density because of divergence of the charge transport by all participating carriers, that is, by ions, cloud droplets, and precipitation. It is easy to show how the compensation period is determined.

2. QUANTITATIVE DERIVATION

If N particles enter through the horizontal unit surface per second and, in the average, gain a charge increment dq, each, while falling through the vertical unit length dz, the cloud air loses the charge $-N\partial\bar{q}/\partial z$ per second and unit volume. The initial space charge density of the precipitation particles falling with a mean velocity $\bar{u}$ relative to the air is $N\bar{q}/\bar{u}$. The neutral state is reached when

$$\frac{N\bar{q}}{\bar{u}} = -N\frac{\partial\bar{q}}{\partial z}t \tag{1}$$

where t = time reckoned from the instant when the precipitation first arrives. The compensation time is then given by

$$t_{\zeta} = \frac{\bar{q}}{\dfrac{d\bar{q}}{dt}} \tag{2}$$

where $d\bar{q}/dt$ is the individual rate of change of the average particle charge. The fact that the compensation time is defined by this kind of ratio suggests that certain parameters, like the charging constant, drop out.

It is not difficult to estimate the order of magnitude of this ratio if we assume charging mechanisms which are either mass or surface dependent. From accretion studies it can be derived that the mass of a precipitation particle grows after an exponential law. It then turns out that

$$t_{\zeta} \cong \frac{m}{\dfrac{dm}{dt}} \tag{3}$$

It can be shown that this law applies to most charging mechanisms one can think of (Kuettner, 1956).

For graupel particles growing by accretion, the ratio is quasi-constant in a wide range of Reynolds numbers or particle sizes and is of the order of 100 seconds.

A similar computation for snow crystals growing by sublimation from the vapor phase shows an identical law, however, the ratio is here proportional to the radius. The compensation time exceeds 500 seconds for crystals of more than 1 mm radius. Fig. 2 illustrates these results.

It is remarkable that the compensation time is independent of the concentration of precipitation particles and of the charging rate and

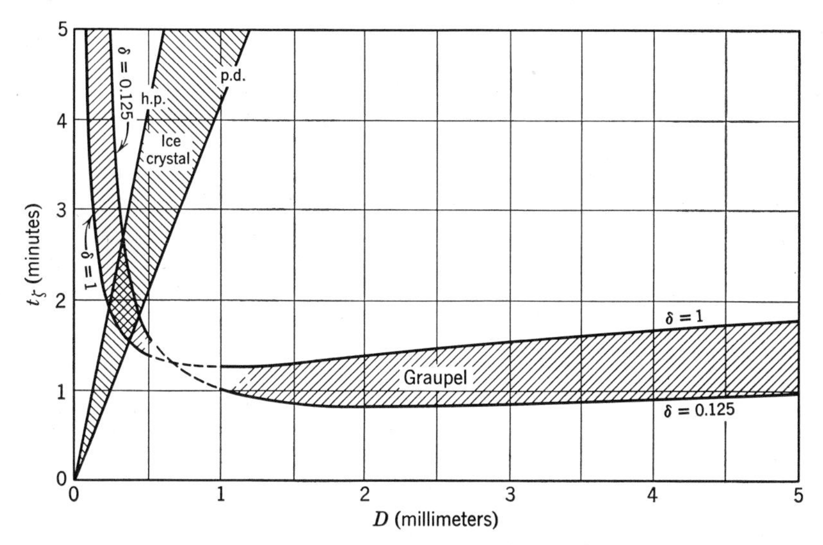

Fig. 2. Compensation period t_ζ versus diameter of precipitation particles (h.p. = hexagonal plates, p.d. = plane dendrites, δ = particle density).

therefore of great generality. Actually the true compensation time is even shorter and, for graupel precipitation, nearer to one minute if air conductivity is taken into account.

The compensation time marks the instant when cloud charge and precipitation charge mask each other completely. As soon as this period is exceeded, the excess of cloud charge over precipitation charge determines the electric field gradient. We have seen that this net charge would grow indefinitely through the precipitation mechanism if the ionic conduction currents set up by the electric field did not counteract this growth and approach an asymptotic state. Eventually the ratio of excess charge to precipitation charge reaches a certain value which we call the masking factor. It must be expected that a high growth rate of precipitation tends to enhance this factor (and to shorten the compensation period), whereas a high conductivity tends to reduce the masking factor (and to shorten the ionic relaxation period).* It is, therefore, not surprising that the masking factor equals the ratio of compensation time to ionic relaxation time. Figure 3 shows the masking factor as a function of air conductivity and precipitation type. Evidently a thunderstorm cloud must be filled with

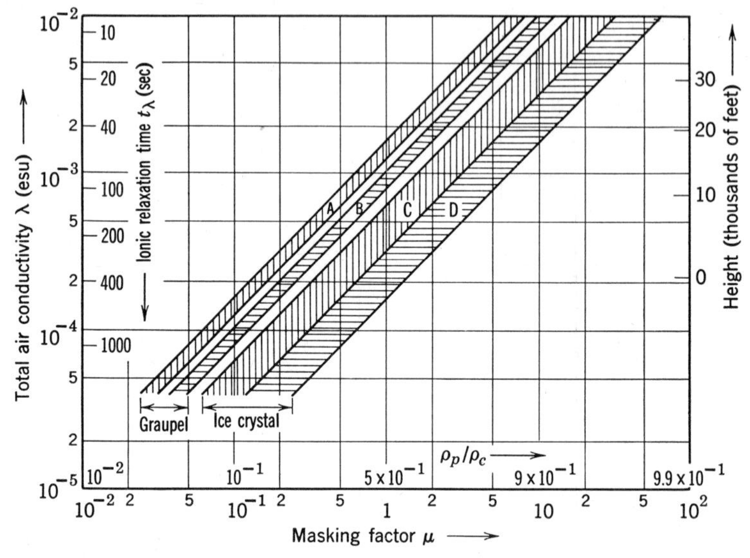

Fig. 3. Masking factor μ versus air conductivity λ for different types of precipitation. A = graupel particles, density 0.125, diameter 1 to 6 mm. B = graupel particles, density 1.0, diameter 1 to 4 mm. C = ice crystals, plane dendritic, diameter 0.5 to 1.0 mm (increasing toward right). D = ice crystals, hexagonal plates, same diameters as C. ρ_p = precipitation space charge (esu/cm^3), ρ_c = cloud space charge (esu/cm^3).

hidden space charges and the masking process must be one of its fundamental properties.

The sudden "unmasking" of these hidden space charges may not only occur at the "stoplevels" and by fluctuations in precipitation intensity but also by lightning discharges.

If precipitation passes a stoplevel, like the freezing level, the masking ratio depends on the polar air conductivities. If they are equal, the precipitation of space charge will exceed the cloud charge by a factor of 2.

* The relaxation period is defined by the time required for a given charge to decay through air conductivity to $1/e$ of its initial value.

The question now arises how the masking process affects the electric field. All one can conclude from this mechanism and from Poisson's law is that, in the equilibrium state, the vertical variation of the precipitation current must be compensated by a corresponding variation of the ionic conduction current. That is

$$\lambda \frac{\partial F}{\partial z} = -\left(\frac{\bar{q}}{m}\right)\frac{\partial I}{\partial z}$$

where λ = total air conductivity (esu), $\bar{q}/m$ = specific precipitation charge (esu g^{-1}), I = precipitation intensity (g cm^{-2} sec^{-1}). As a consequence the equilibrium condition does not determine the electric field itself but its vertical variation only. The magnitude of the field depends entirely on the boundary conditions and may change with them. This is in contrast to Wilson's electric-gravitational equilibrium.

A numerical example may be of interest. If the precipitation intensity increases over a certain vertical distance by an amount of about 1 inch per hour or 10^{-3} (g cm^{-2} sec^{-1}) and the air conductivity is 10^{-3} esu, the variation of the electrical field over the same distance is (numerically) of the same order as the specific charge $\bar{q}/m$. Thus a mean precipitation charge of $+5$ esu per gram of ice will correspond to an electric field increase of 5 esu or 1500 volts per cm. As the sign shows, this field increase is directed upward, just as in nature, but opposite to what should be expected from the positive polarity of the precipitation charge. This is a consequence of the masking process which has created a more intense cloud charge of opposite polarity. This result seems to be in agreement with the findings in nature as Fig. 4 shows.

3. CONCLUSIONS

We may now summarize the situation: Meteorological factors enter the problem of charge distribution in thunderstorms. In the mature storm vertical motions are directed upwards in the upper cloud portion and of sufficient magnitude to overcome the falling speed of most of the precipitation. In the lower portion vertical motions are mostly downwards, especially in the center of heavy precipitation. The two main charges are seated in these two portions of the thundercloud and at temperatures well below freezing.

Observational evidence points to a close relation between charge

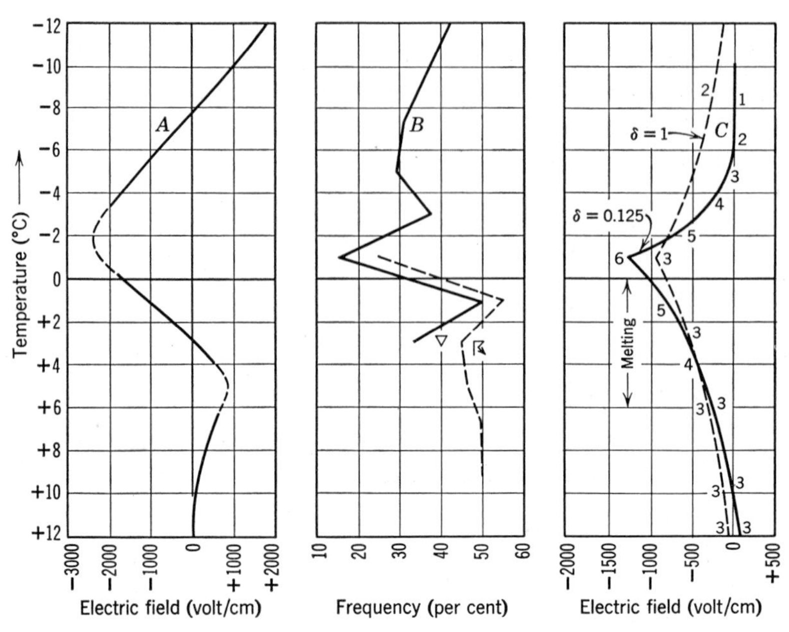

Fig. 4. Observed and theoretical types of electric fields in thunderstorms. Vertical field component as function of temperature. A = thunderstorm model according to measurements of Simpson and Scrase (1937) and Simpson and Robinson (1941). B = mountain observations at Zugspitze (Kuettner, 1950) based on 118 thunderstorms and showers; solid curve: shower clouds without lightning; dashed curve: thunderstorms. C = theoretical curve for positively charged graupel particles at "basic charging rate" of +5 esu per gram of ice; numbers along curves denote particle diameters in millimeters; δ = particle density.

generation and the development of solid precipitation in the supercooled cloud. In fact, the appearance of the ice anvil has, for a long time, been considered, by meteorologists, as the identifying mark of a thunderstorm.

The peculiarity of the charge segregation process now leads to a masking of precipitation and cloud space charges in such a manner that the cloud charge slightly dominates over the opposite precipitation charge in the downdraft portion while the opposite is true in the updraft portion. In this way a vertical charge dipole is created in the mature thundercloud.

"Unmasking" of the hidden charges occurs at temperature levels where the growth of precipitation stops or changes. One of these so-called "stoplevels" is the freezing level beneath which the precipitation charge is partly unmasked.

The average charge distribution found in thunderstorms places the positive main charge into the updraft portion at temperatures around $-30°$ C, the negative main charge into the downdraft portion at temperatures between $-5°$ C and $-10°$ C, and Simpson's lower positive charge center into the freezing "stoplevel." From this it should be concluded that the two positive charges are carried by precipitation, the negative charge by cloud droplets. The polarity of the solid shower precipitation, then, should be positive in contrast to what has generally been assumed but in agreement with many observations.

As another consequence of the masking process, fluctuations in precipitation intensity may unmask hidden charges and create strong local field gradients.*

There are several implications which are of significance to other problems. One of them concerns the electric budget and the total current of a thunderstorm. Another is the short time factor of the field recovery curve which seems to postulate a very high conductivity throughout the thundercloud after a lightning discharge. Finally there is the question of how inadequate our vertical one-dimensional models are and how important horizontal gradients may be.

The last two points gain interest in the light of a new achievement of radar meteorology. Through the pioneer work of Dr. M. G. H. Ligda (formerly with the Meteorology Department of M.I.T.), it has been possible to make lightning processes visible on the radar screen. This may open an entirely new field of thunderstorm research. I am sure that the short motion picture which has resulted from this work would have fascinated Benjamin Franklin. (See Fig. 9.)

As a comment on the problems raised by this film and the masking process I would like to mention a phenomenon which is frequently observed after a lightning discharge but has received little attention. (See Figs. 5 and 6.)

Immediately after the lightning stroke the electric field may be a multiple of the predischarge field but of opposite sign. This high post-discharge field is the one which decays in the recovery curve with a time factor of about 7 to 10 seconds. The event is so common that, for example, in the mountains the approach of a thunderstorm is first indicated by numerous, short-lived point discharges which give a hissing noise at very distant lightning. This happens even before thunder is audible. Franklin had observed this phenomenon; in his letter to Collinson of September 1753 he reports: "I found . . . that some-

* For further details, see Kuettner, 1956.

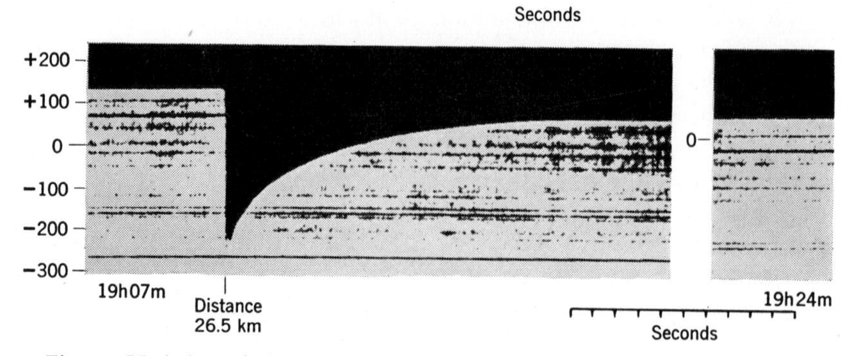

Fig. 5. Variation of electric field during distant lightning (Wormell, 1939).

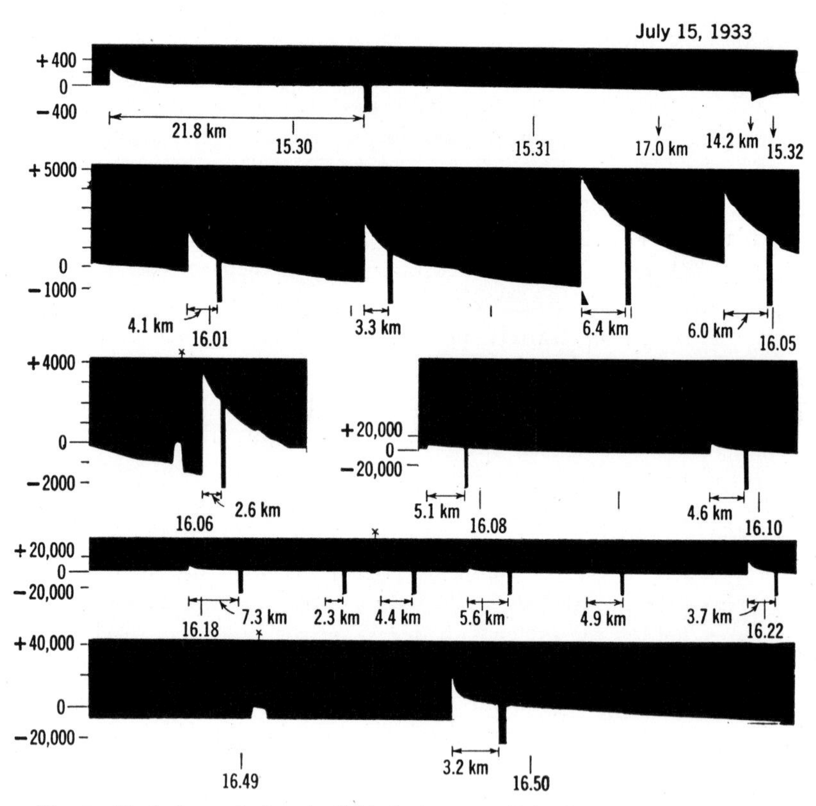

Fig. 6. Variations of electric field during near lightning (Wormell, 1939).

times after a flash of lightning the bells would suddenly stop; and at other times when they had not rung before, they would, after a flash, suddenly begin to ring. . . ." Frequent lightning discharges may deceive an observer by suggesting an electric field of the wrong polarity which is actually only a sequence of reversed postdischarge fields.

Fig. 7 gives an example of a normal field record with a few lightning discharges while Fig. 8 shows how the equilibrium field is overshadowed by repeated postdischarge fields. In this case lightning is so frequent that it does not give the electric field at the point of the observer enough time to recover to its original polarity.

For an untrained observer this game of nature resembles a sort of jack-in-the-box. Hidden charges seem to pop up after the lightning discharge only to be hidden again until the next lightning strikes. Is it possible that masked charges are exposed by lightning because their counterparts are discharged? An interpretation along this line has been given by Wormell (1953) who suggested a space charge blanket near the ground which masks the cloud charges and is exposed after their discharge. This explanation does not satisfy the conditions inside the thundercloud and on well-ventilated mountains. A continuous internal masking appears more appropriate and fits into the masking mechanism. The question then arises whether the lightning discharge does not affect the electric equilibrium over wide areas, the strong postdischarge fields giving rise to numerous corona discharges from particles whose charge densities have been below but close to the breakdown limit. This could provide a sudden extraordinary increase in air conductivity as indicated by the short time factor of the recovery curve. How the charge drainage from such vast cloud volumes can be accomplished is the subject of Professor Loeb's essay in this volume. That areas of unexpected dimensions are affected by lightning will be obvious from Fig. 9. It may not be correct to identify lightning echoes which flash over horizontal distances of 30 to 100 miles with visible lightning channels; they may be merely areas of high ionization. Also the empty spaces over which the echoes spread may not be clear sky but clouds, since the radar sets used depict only precipitation areas. But the picture leaves no doubt that the horizontal dimensions of space charges involved in thunderstorm discharges are tremendous. The lack of lightning echoes at close distances suggests that the height of the lightning activity is over 30,000 ft where they are not caught by the low radar beam.

I hope that a grain of truth may exist in the presented masking

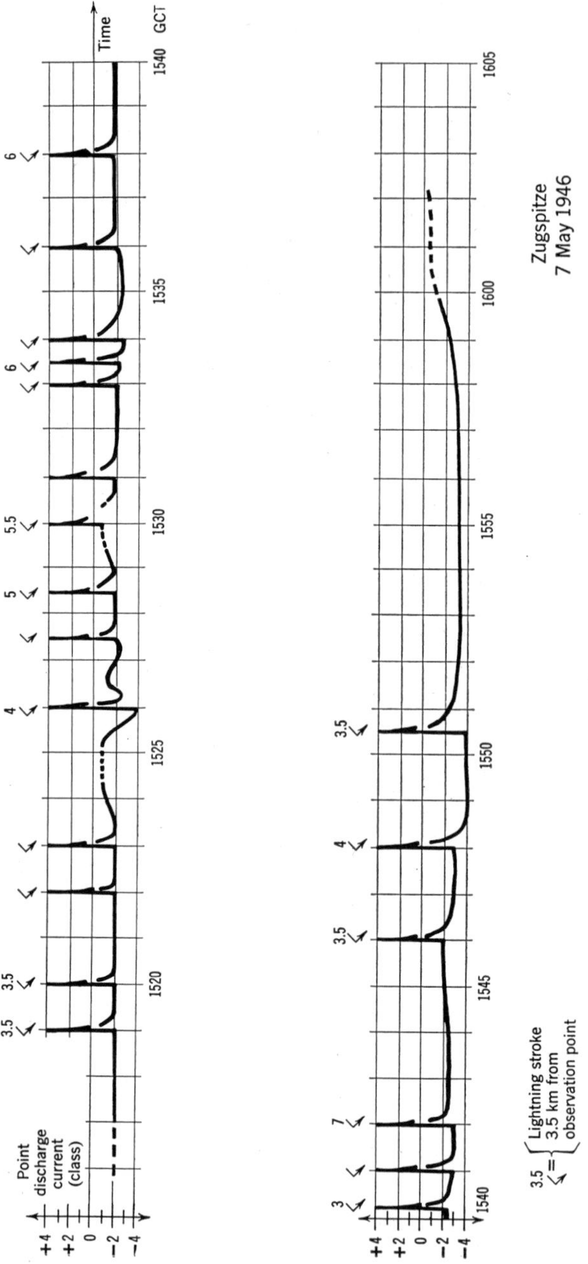

Fig. 7. Reversals of corona current by lightning strokes from a normal thunderstorm, as measured at the 10,000-ft Zugspitze Observatory (Germany).

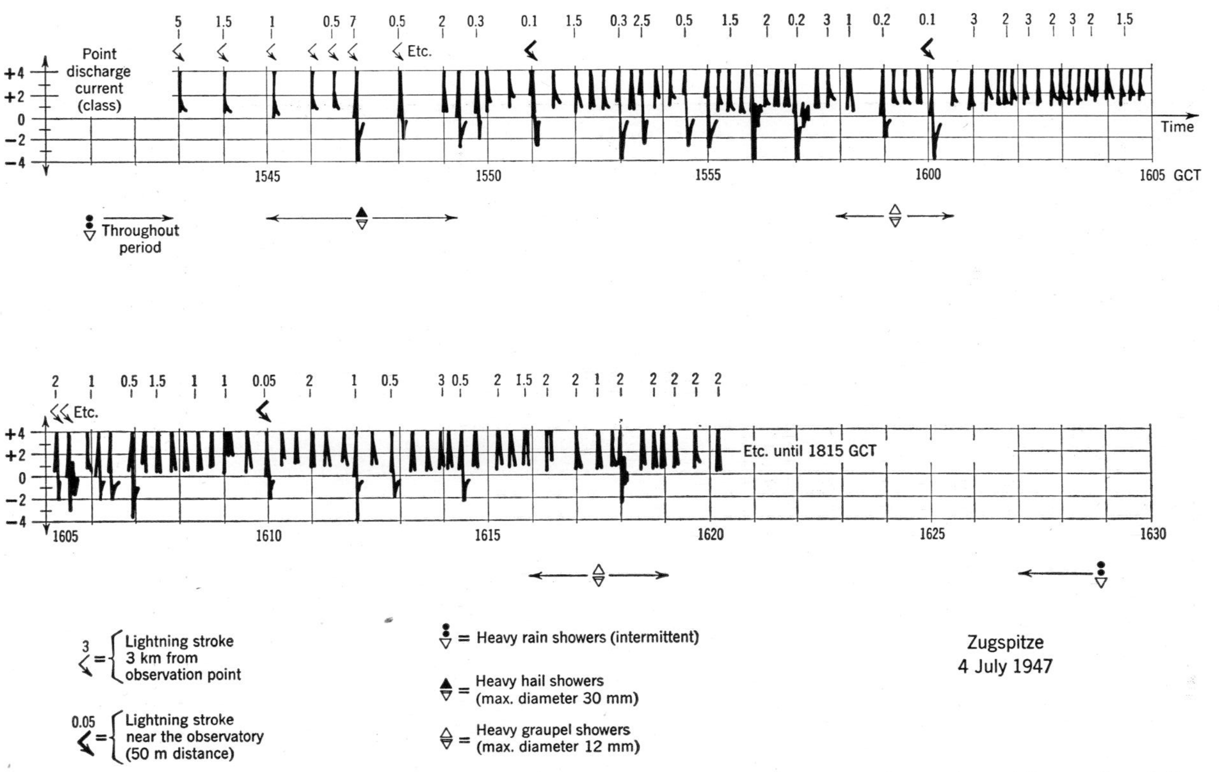

Fig. 8. Reversals of corona current by lightning from a severe thunderstorm over the station. The strokes follow each other at such a high rate that the field has no chance to recover, thereby indicating an average field of reversed sign.

Fig. 9. Radar photograph of lightning close to 100 miles in horizontal length, flashing from the rear of a squall line in midwestern United States. The two pictures are consecutive frames, the upper one showing the lightning echo (23-cm radar, 100-mi range setting) (Ligda, 1956).

theory but also concede that now, 200 years after Franklin's work, our understanding of this impressive display of nature is still inadequate.

REFERENCES

Byers, H., and R. Braham, 1949, *The Thunderstorm*, U. S. Weather Bureau, Washington, D. C.

Cohen, I. B., 1941, *Benjamin Franklin's Experiments*, Harvard University Press, Cambridge.

Findeisen, W., 1940, Über die Entstehung der Gewitterelektrizität, *Z. Meteorol.*, **57**, 201–215.

Gunn, R., 1948, Electric Field Intensity Inside Natural Clouds, *J. Appl. Phys.*, **19**, 481–484.

———, 1950, The Free Electric Charge on Precipitation Inside an Active Thunderstorm, *J. Geophys. Research*, **55**, 171–178.

Israel, H., H. Kasemir, and K. Wienert, 1955, Luftelektrische Tagesgänge und Massenaustausch im Hochgebirge der Alpen, *Arch. Meteorol. Geophys. Bioklimatol.*, **8**, 72–94.

Kuettner, J., 1950, The Electrical and Meteorological Conditions Inside Thunderclouds, *J. Meteorol.*, **7**, 322–332.

———, 1956, The Development and Masking of Charges in Thunderstorms, *J. Meteorol.*, **13**, 456–470.

Ligda, M., 1956, The Radar Observation of Lightning, *J. Atm. and Terrest. Phys.*, **9**, 329–346.

Lueder, H., 1951, Ein neuer elektrischer Effekt bei der Eisbildung durch Vergraupelung in natürlichen unterkühlten Nebeln, *Z. angew. Phys.*, **3**, 247–253 and 288–295.

Schaefer, V., 1947, Properties of Single Particles of Snow and the Electrical Effects They Produce in Storms, *Trans. Am. Geophys. Union*, **28**, 587.

Simpson, G., and F. Scrase, 1937, Distribution of Electricity in Thunderclouds, *Proc. Roy. Soc. (London)*, **A 161**, 309–352.

Simpson, G., and G. Robinson, 1941, The Distribution of Electricity in Thunderclouds II, *Proc. Roy. Soc. (London)*, **A 177**, 281–329.

Wilson, C. T. R., 1929, Some Thundercloud Problems, *J. Franklin Inst.*, **208**, 1–12.

Wormell, T., 1939, The Effects of Thunder Storms and Lightning Discharges on the Earth's Electric Field, *Phil. Trans. Roy. Soc. (London)*, **A238**, 249–303.

———, 1953, Atmospheric Electricity; Some Recent Trends and Problems, *Quart. J. Roy. Meteorol. Soc.*, **79**, 3–38.

LEONARD B. LOEB
Professor of Physics
The University of California,
Berkeley

The Positive Streamer Spark in Air in Relation to the Lightning Stroke

Ever since the kite string experiment of Benjamin Franklin carried an electrical charge from the thundercloud, stored it in a condenser, and yielded an ordinary electrical spark, the dreaded thunderbolts of Jupiter have been identified in numerous ways with the electrical spark on however a grandiose scale. As techniques improved over the years, similarities and differences have appeared. What has primarily emerged is that both phenomena are exceedingly complex as we shall see.

Probably the greatest advance in the understanding of the lightning stroke came in the work of B. F. J. Schonland [1] and his colleagues through the use of the rotating lens camera invented by C. V. Boys. Fig. 1 presents one of the most striking of these photographs showing the lightning stroke descending from a negatively charged cloud cell in a series of luminous *steps* at an average speed of advance of about 10^7 cm per sec. On junction of the last step with the earth, the conducting channel forged is reilluminated by the brilliant *return stroke* characterizing the brightly luminous flash which we see. Actually, as indicated in Fig. 1, this return stroke represents a *sequence* of *bright flashes* advancing from ground to cloud at intervals of some

10^{-2} sec with speeds which, in some instances, lie around 10^9 to 10^{10} cm per sec.

Many studies [1] have confirmed that *most* lightning strokes to earth come from the negatively charged cloud cells several kilometers in diameter and extending in height above the earth from some 3 to 9 km. In the cells, the charge is stored on a small fraction of the myriads of raindrops. It generally is assumed that the stepped leader which is shown in Fig. 1 is preceded by an invisible *pilot leader* that advances some 10 to 160 m, as indicated by the length of the steps depending on the vigor of the stroke, into virgin air toward the ground on a sort of meandering path. After such an advance, decay of ionization and conductivity in the pilot leader channel lead to an accumulation of electricity and an increase of electrical field at the cloud end of its path, which, at an appropriate value, launches a wave of ionization and luminosity down to the tip of the then slowing pilot leader. This ionizing wave moves at from 1.5×10^7 to 1.5×10^8 cm per sec and increases conductivity and luminosity of the pilot leader by a factor of perhaps a thousand. It shows a breadth which may be considerably greater than that of the pilot leader. This brilliant manifestation constitutes the luminous step observed in Fig. 1 and revitalizes the tip of the pilot leader that again forges ahead as before. The steps are relatively regular in length in any one stroke, but vary widely in different strokes.

When the bright stepped leader approaches within some 30 to 40 m from the ground, a so-called positive streamer is initiated from the ground by the enormous electrical field distortion produced. This streamer may advance 10 to 20 m before it connects with the stepped leader. The distortion of the electric field at the junction point of the two produces the electrical field distortion that propagates the ionization and creates the high luminosity of the return stroke.

While the phenomena observed are in some measure logically interpreted in terms of the observation, much more information is needed to understand in detail the character of positive and negative streamer processes and more still to understand how the electrical charge distributed over the myriad of charged droplets in the large volume ($\sim$1 km^3) of cloud cell can be gathered together and fed into the streamer during the 0.01 second (approximately) involved in the single stroke. To assist in clarifying these processes, we may profitably turn to recent advances in the understanding of the mechanism of the common crooked electric spark in air with which we are all familiar.

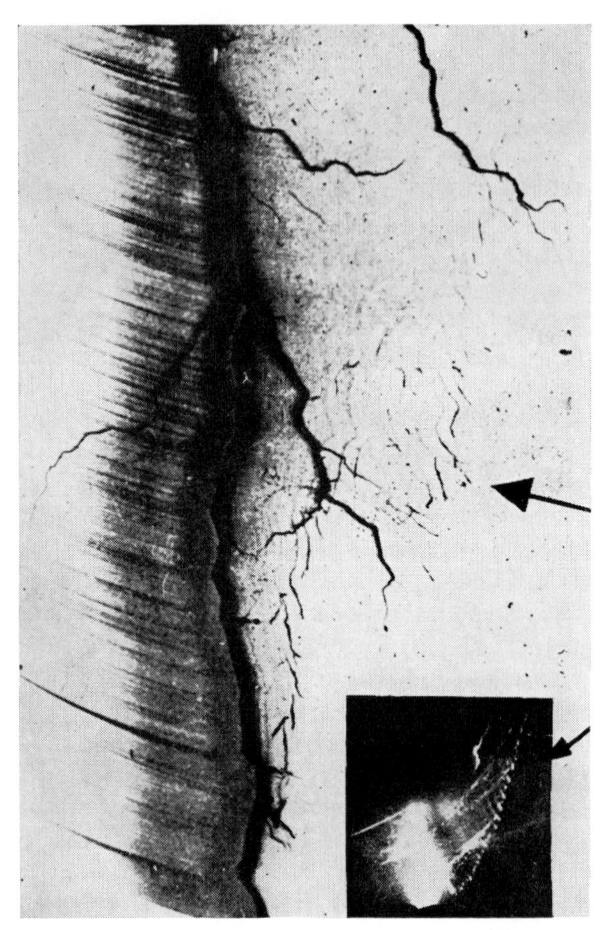

Fig. 1. Schonland's rotating lens, Boys camera pictures of two typical stepped leaders and return strokes. Rotation and time displacement from right to left. Stroke advances downward. Note the successive steps moving downward diagonally at arrows on the right. The dark line is the main or return stroke which also reilluminates the downward branches. A second and perhaps a weaker third stroke follows to the left, somewhat obscured by the persistence of luminosity. Inset lower right shows another stroke; in this case the photographic print is a positive.

1. STREAMERS IN IMPULSE BREAKDOWN

In 1935–1936, H. Raether and E. Flegler [2] in Germany studied sparks produced in plane parallel gaps when an impulse overvoltage of square wave form of known duration was applied inside a C. T. R. Wilson cloud chamber. Ultraviolet light liberated electrons from the negative electrode. The high potential caused these initial electrons to multiply while in transit to the positive electrode by cumulative ionization, producing what is called an electron avalanche. By stopping the high field abruptly while the electrons were in mid-gap, or at later times and then adiabatically expanding the air saturated with alcohol and water vapor, cloudy condensation occurred on the ions of the avalanche. Illumination of the cloud track of condensed ions from the side allows the track to be photographed. Figs. 2 and 3 show such avalanches photographed by Raether.[3] By varying the time of duration of the pulse, the velocity of avalanche advance was measured and found to lie around 2×10^7 cm per sec. At higher fields, the avalanche having arrived at the positive electrode, or in still higher fields, having arrived at mid-gap, underwent a sudden change.

If it arrived at the positive plate, it sent a positive streamer back to the negative plate. At still higher fields, in mid-gap, it assumed a spindle shape, sending a positive streamer to negative plate and a negative streamer to anode.[4] These are shown in Fig. 4. Early in 1936, the writer and his students [5] quite independently discovered the positive streamers proceeding from a positive needle point in corona discharge in air and suggested that these streamers led to the common spark. Both Raether and Flegler and the writer independently interpreted the streamer mechanism process in the same fashion. Figs. 5 and 6 are schematic diagrams illustrating the principle of streamer formation. The writer went further than Raether and Flegler in that on the analogy of Schonland's lightning discharge observations, he attributed the bright phase of the spark not to an increasing intensity in the streamer current in time, but to a *return stroke.*

In 1938, T. E. Allibone and J. M. Meek,[6] initially in cooperation with Schonland, took moving film photographs of meter-long streamer sparks on impulse breakdown from a point to a plane in air. Figs. 7 and 8 show these. It may be noted that with series resistance stepping can be observed in the positive point streamer. Here streamer and return stroke are clearly seen. With negative point and conduct-

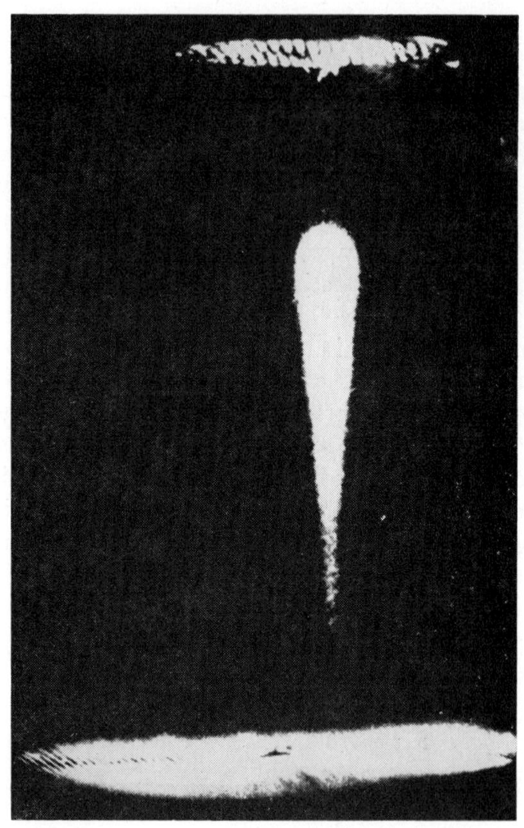

Fig. 2. Raether's Wilson-type cloud-track photographs of water droplets condensed on ions left by an electron avalanche in a high field impressed for fractions of a microsecond in nitrogen near atmospheric pressure much enlarged. The negative electrode is at the bottom, the single initiating electron being released from it by ultraviolet light. Reduction in expansion ratio and re-evaporation of the droplets fail to record the few individual drops near the origin. The avalanche has advanced some two thirds of its distance to the wire anode at the top.

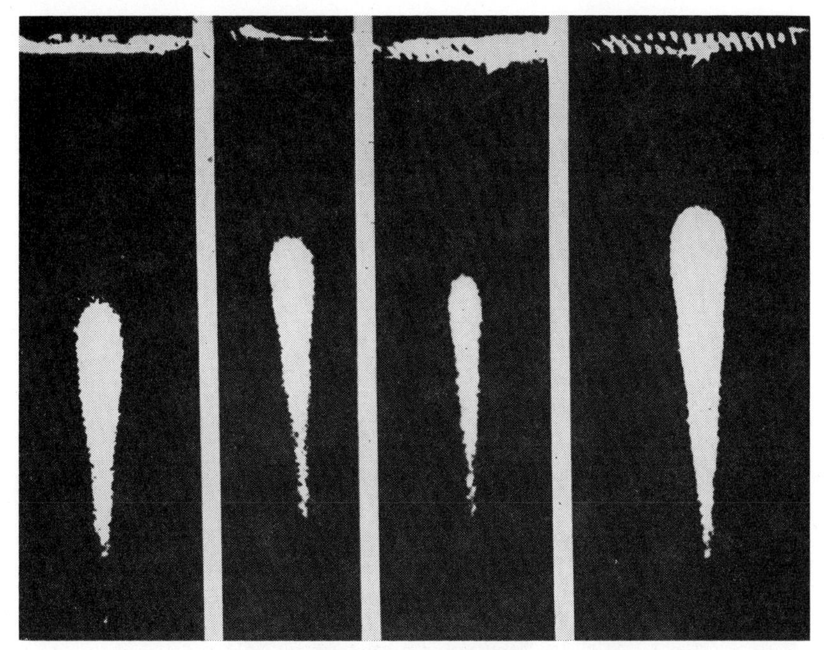

Fig. 3. Tracks similar to those in Fig. 2, showing avalanches in different gases and indicating how the distance of advance measured for different durations of the high field permits drift velocities to be measured.

ing positive plane, the negative streamers are seen to have difficulty in starting. When they start to progress, they need twice the potential required for positive points. The negative streamers advance only ⅓ of the gap length before the high field at the positive plane initiates a fast positive streamer that meets the negative streamer in midgap. From the junction point, the return strokes proceed *up* and *down* to the electrodes. The predominance of the positive streamer advance over that of the negative streamer stems from the conservative nature of the positive streamer in that it draws the nimble electrons into its nearly immobile positive charge, instead of outwardly dispersing them as from a negative cloud. It is thus logical to consider the more rapid advance of positive streamers relative to the negative streamer process as the important factor in cloud drainage in the negative stepped lightning stroke advance. It should be noted in passing that the lightning stroke partakes much more of the nature of an impulse breakdown than of the static breakdown to be discussed in order to delineate more precisely the character of the process.

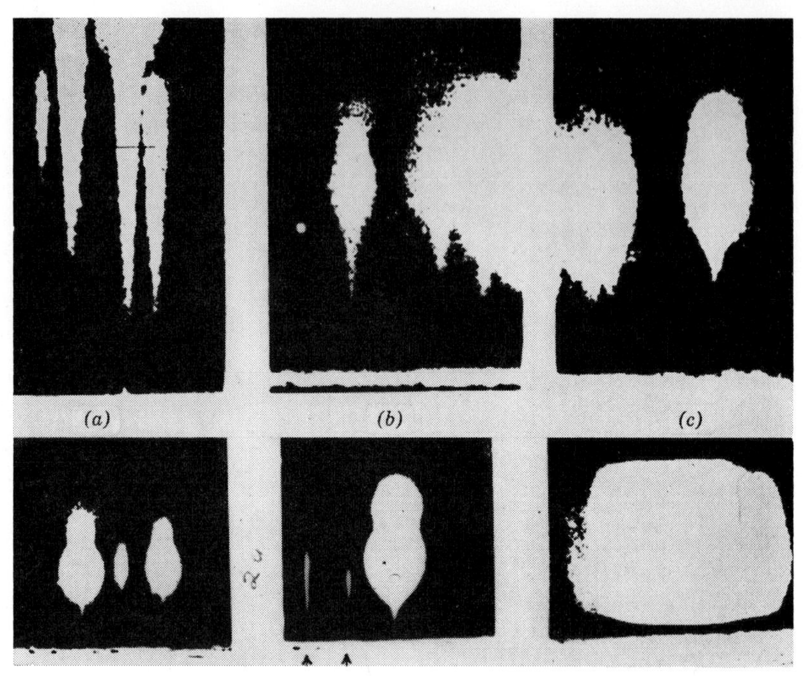

Fig. 4. The transition of the avalanche to a streamer above sparking fields as revealed by Raether's cloud-track pictures. Cathode below, anode above. In the upper row exposure (*a*) is at spark threshold. Several avalanches cross. The third from the left reaches the anode ahead of the others and the bulge shows the beginning of a positive anode streamer. Exposures (*b*) and (*c*) show the development of the spindle indicating the creation of mid-gap *anode* and *cathode* streamers at considerable *overvoltage*. Note the presence of avalanches that started later in several exposures. Note particularly the greater development of the lower positive cathode-directed streamer (below), relative to the negative anode-directed streamer above. The former moves more rapidly than the latter. The lower right exposure shows the cloud track of the completed and expanded spark channel created by the return stroke.

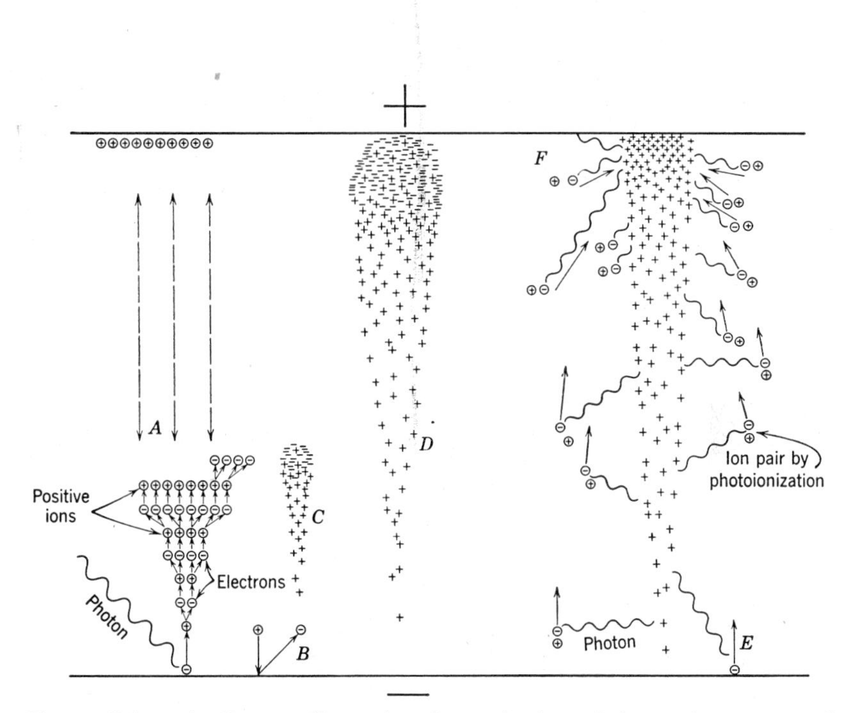

Fig. 5. Schematic diagrams illustrating the mechanism of the anode streamer of Fig. 4*a*. The sequence starts at the left with the growth of an avalanche created by a single photoelectron liberated from the cathode below at *A*. *C* shows an avalanche in its early stages. *D* shows an avalanche as it crosses the gap to the anode. *F* shows the electrons swallowed by the anode, leaving a positive space charge to increase the anode field. The wavy lines indicate photons creating electrons *in the gas* which produce new avalanches feeding into the space charge and advancing it toward the cathode.

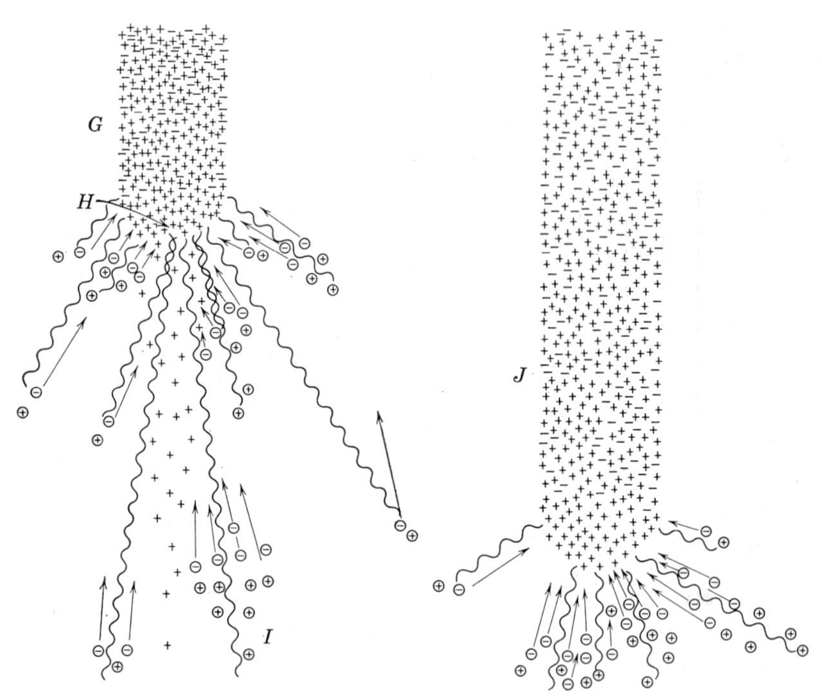

Fig. 6. A continuation of the sequence in Fig. 5, showing the advance of the cathode-directed positive or anode streamer downward across the gap to the cathode. The channel *G* is conducting in consequence of electrons feeding in laterally (not shown) and from the tip. At *H* the positive space charge presents a very high tip field, causing the advance. In Hudson's oscillograms shown in Fig. 10 the intense ionization near the advancing tip produces the bright pip of light called the *primary* streamer, while the lateral spread and increased electron flow up the channel some distance above the tip, as at *J*, mark the *secondary* luminous streamer. This is not indicated in the diagram at *J*. The intense electron emission from the cathode as a result of photo-ionization and field distortion as the tip approaches the cathode indicated in *J* causes the cathode flashes of Fig. 14 of Hudson.

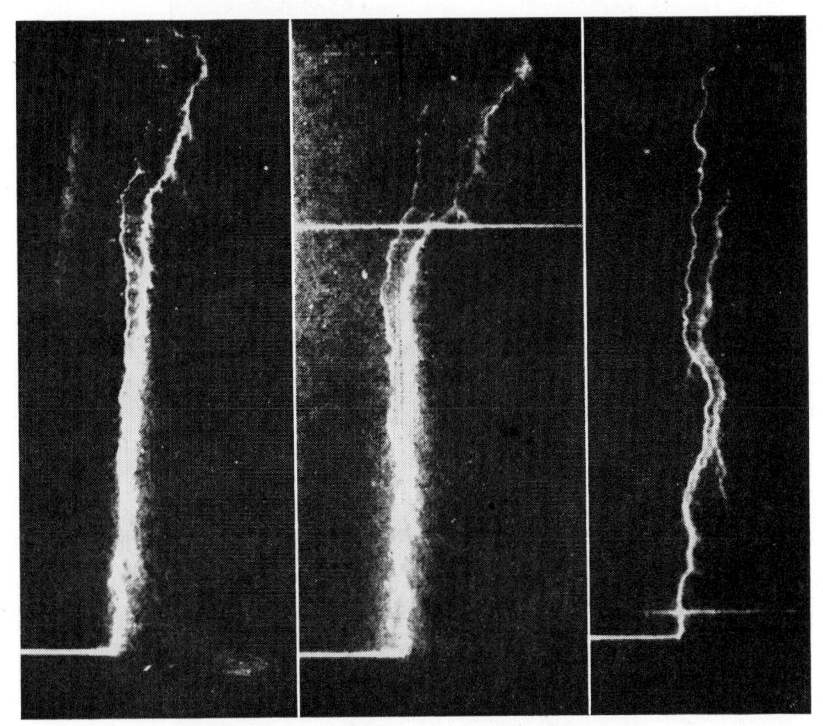

Fig. 7. Allibone and Meek's moving film camera photographs of meter-long sparks in air from positive point (top) to plane (bottom). The film and time progress from right to left, the streamer advancing downward diagonally. The "return stroke" on the left moves so rapidly, $> 3 \times 10^8$ cm per sec, its speed cannot be measured. The central exposure was made with high series resistance, and the stepping resulting can barely be discerned near the top. These are the first indication of the return-stroke process in common sparks which appears to be confirmed by Hudson for the shorter, less impulsive common sparks at the threshold.

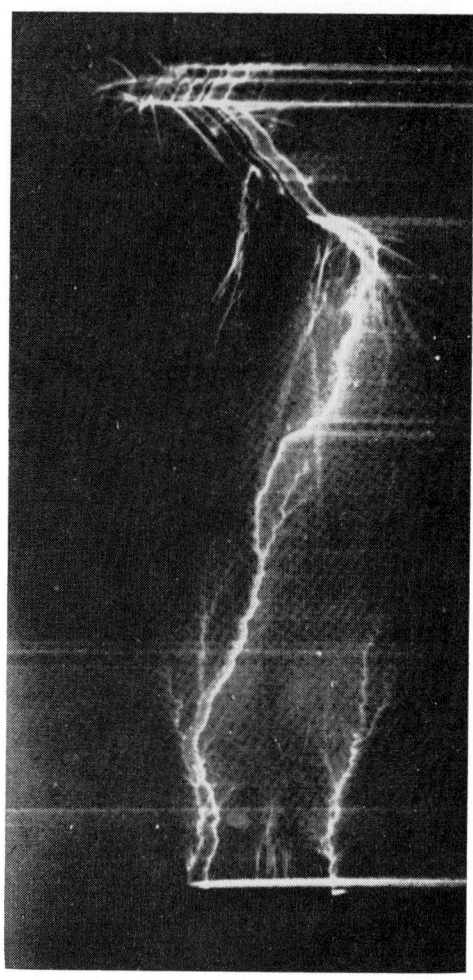

Fig. 8. Allibone and Meek's photograph of an impulse spark from a negative point (top) to positive plane below. Here time and film move from left to right. The impulse potential is nearly twice that for a positive point to plane gap of the same length. Note the successive unsuccessful steps of the negative streamer and the development of a positive streamer from the anode at the left when the negative step has progressed about one third across the gap. The positive and negative streamers meet at that point and return strokes move back to the anode and cathode from the junction point. Note that, aside from the small displacement to the right because of film motion, the spark in itself is crooked because of the origin of the successful positive streamer. Note other positive streamers starting from the plane at a later time. Had one of these started earlier, the spark would not have been so crooked.

2. STREAMERS IN STATIC BREAKDOWN

Although the studies of Raether and Allibone and Meek left little question as to the streamer character of the impulse spark breakdown, time resolution was then not adequate to establish details and the streamer mechanism was not established for the common static spark. Proof of the presence of streamers as a factor in uniform field sparks became more urgent when, in 1951, L. H. Fisher and B. Bederson [7] showed that uniform field breakdown was *preceded* by a low order diffuse glow discharge that, through field distortion, led to a spark breakdown into a filamentary crooked streamer channel. Recently, G. G. Hudson,[8] in the writer's laboratory, has established the positive streamer to be the primary mechanism in static spark breakdown in room air at 760 mm. Gaps ranged from 1 to 5 cm in length, with positive electrodes ranging from small needle points to 10-cm diameter spheres against a large plane negative electrode. More recently sparks from 2 to 9 cm long between still larger spheres and plane from a Van de Graaff generator have been studied. His experimental arrangements for smaller gaps are shown in Fig. 9. Fig. 10 presents observations for a small point where fields are not quite equal to breakdown showing a sequence of superposed streamers. Among other things, the remarkable reproducibility of the streamer in the first 1.50 cm of path can clearly be seen. Note the fast tip, called the *primary* streamer, separating from the more slowly advancing increase of conductivity and luminosity by radial ionization and expansion resulting from inward flow of electrons, designated perhaps incorrectly as the *secondary* streamer. Fig. 11 shows the primary-secondary sequence for a small point and long gap at the breakdown, with slow sweep. Here a series of primary-secondary sequences (each sequence not being resolved), separated by some 100 microseconds in time, occurs before breakdown. Fig. 12 shows the same sequence at a sweep so fast that even the return stroke is not shown. Note the great reproducibility of the primary and the differences in the important secondaries, the crossing of which to the plane leads to the return stroke.

Fig. 13 shows the case of a shorter gap and larger point. Here only two primary-secondary sequences precede the return stroke. Note that luminosity of the second successful secondary, as it nears the cathode, rides up on that of the return stroke, time resolution ($\sim 7 \times 10^{-9}$ sec) being inadequate to separate the two. Fig. 14 is a photograph of a 0.4-cm diameter point that yielded a set of pre-

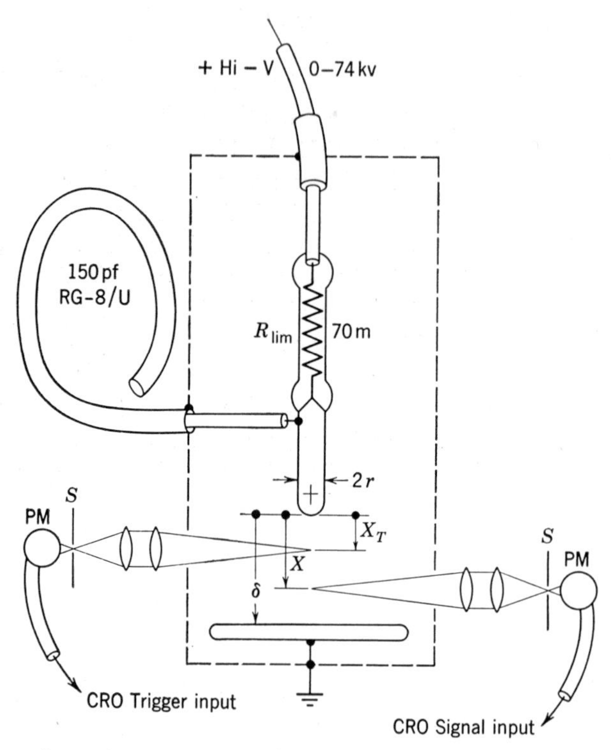

Fig. 9. Experimental arrangements of Hudson for observing the transient pulses of luminosity accompanying a streamer spark sequence using a steady high (70-kv) potential source. Note the limiting resistor R_{lim} protecting the high-tension set, the 150 pf RG-8/U high-tension storage condenser, the positive point and negative plane separated δ cm, and two photomultipliers, the upper one usually scanning the gap very close to the anode used to trigger the sweep and the second one placed at x cm from the point.

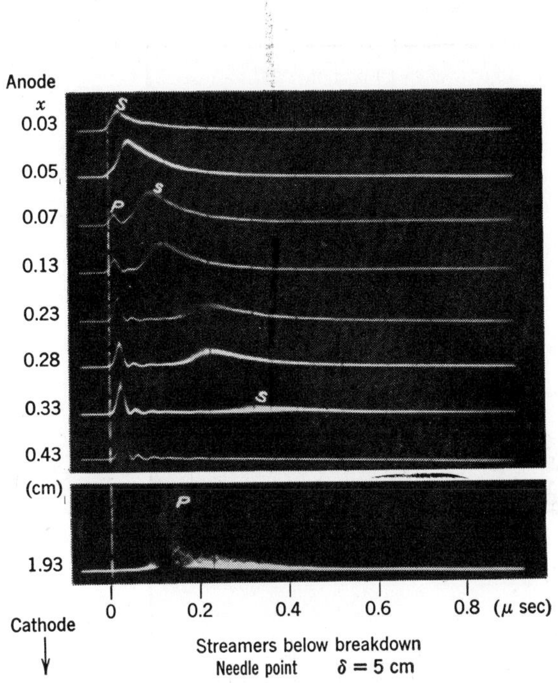

Fig. 10. Observations of Amin and Hudson in the *pre*-breakdown range using a small point, 5-cm gap and fast sweep. The distance x in cm is indicated at the left. The triggering photomultiplier is set 0.01 cm from the point. Note that near the point there is just one streamer pulse which at 0.07 cm divides into two, a faster one labeled p, the *primary* rapidly advancing bright narrow streamer tip, and s, the secondary spatially and temporally more extensive and initially of higher luminosity *secondary*. This represents some 10 or more successive streamers, showing the extreme initial reproducibility of the phenomenon. As the primary advances, it is increasing in intensity probably owing to field distortion ahead of its conducting channel. The secondary created by lateral inflow of electrons encounters positive space charge from the corona in the lower field region. It attenuates and ceases to be visible to the photomultiplier after 0.33 cm. The oscillations between are probably of instrumental origin. Note how the accidental fluctuations diffuse the primary at 1.93 cm. These streamers do not cross the long low space charge fouled region and yield no spark.

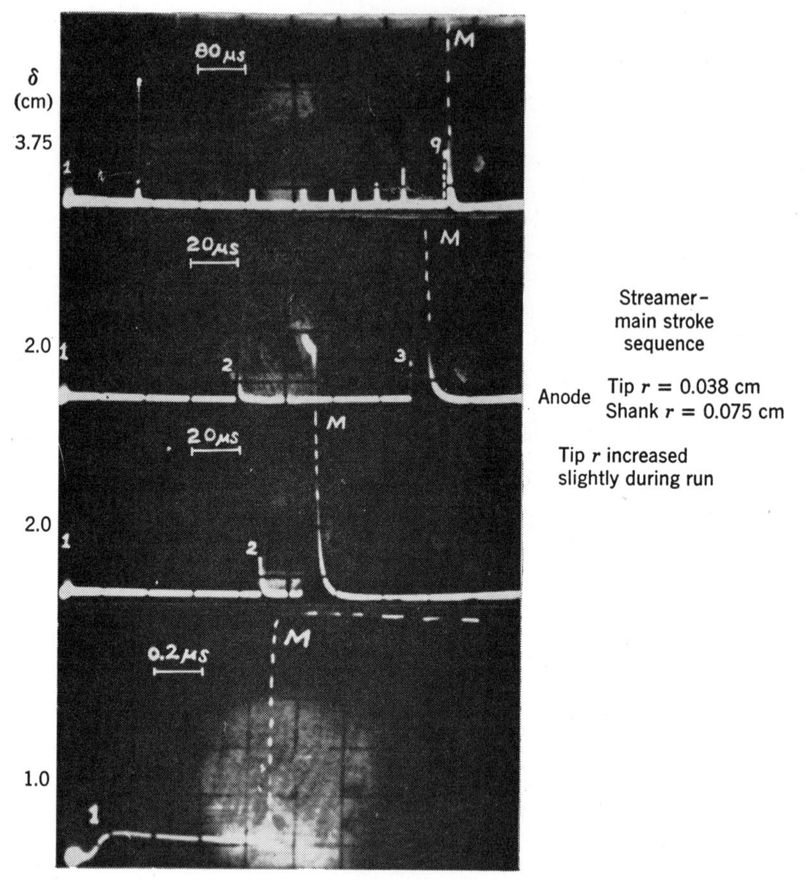

Fig. 11. The breakdown streamer sequence with slow sweep, as indicated by the time scale in microseconds for varying gap lengths and a single point radius of 0.038 cm. With the long gap, there are nine unresolved primary-secondary sequences before a spark appears. The spark is indicated by the rapid rise at *M*. Here the primary and secondary cross the gap from anode to cathode, giving rise to the main or return stroke of such light intensity that it goes off scale at the top. With 2.0 cm only two or three sequences are noted, and at 1 cm the single sequence primary-secondary is followed by a return stroke and spark within 0.4 μsec after crossing has been recorded.

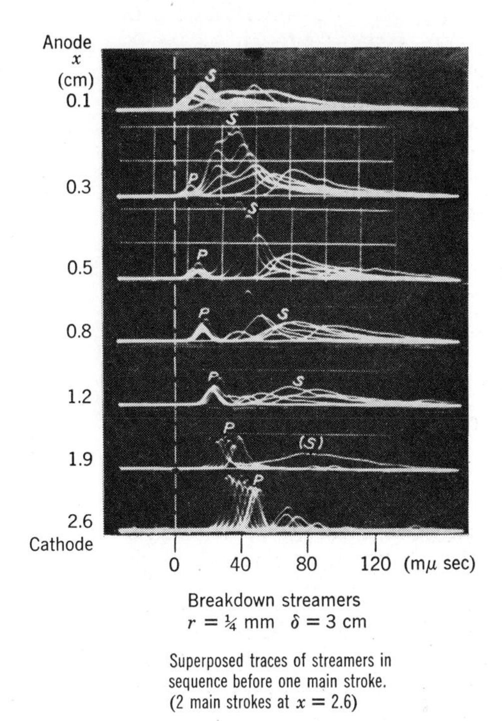

Fig. 12. A superposed sequence of some seven breakdown streamers for a 0.025-cm radius point with 3-cm gap with very fast sweep which excludes the two main return strokes resulting at the right. Note the reproducibility of the primaries in speed and amplitude and the great variation of the secondaries as they cross. This indicates why all secondaries do not yield return strokes. At 2.6 cm there is a delay of some 10^{-6} sec (μsec) between the time of the peak of the successful secondary at 1.9 cm and its arrival and rise of the return stroke (not shown) at the right. This is the time, in principle, required for the slow secondary to traverse the last 0.7 cm of the gap.

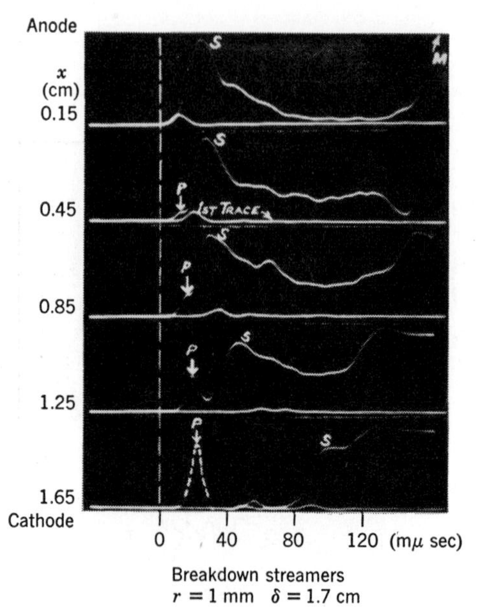

Fig. 13. Breakdown streamer from a point of 0.1-cm radius and a 1.7-cm gap in air with sweep so fast that the main stroke is not fully resolved. In traces at 0.15 and 0.45, one antecedent abortive streamer is seen labeled 1st trace. The primary begins to separate at 0.45 cm, and the beginning of the rise of the main stroke is seen at the extreme right. It rises little, since a cutoff is created by overloading the amplifier. At 1.65 cm the primary is shown just before it arrives at the cathode. The secondary crosses and rises at 80×10^{-9} sec, riding up on the foot of the main stroke.

breakdown streamers near, but below, the breakdown potential. Here, visually and photographically, it can be seen that, after traversing some 8 mm along the axis in a 3-cm gap, the primary and secondary streamers show a tendency to branch. The eye can follow the branching of the *primaries* in a single pulse across the gap. The camera is not as sensitive as the photomultiplier or even the eye, so that a group of forty sequences is shown. Note the bright flashes at the negative electrode where the primary streamer tips sharply, distorting the cathode field strike. The secondaries branch for some 0.8 cm beyond the axial sections of the paths and then fade by attenuation with excessive branching. However, when a fortunate one of these secondaries reaches the cathode, the bright crooked return stroke of the type shown to the left of Fig. 14 occurs.

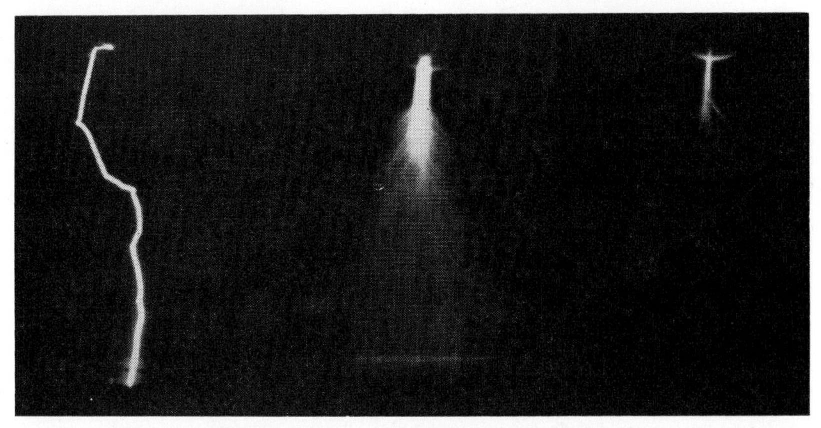

Fig. 14. Still photograph of 0.4-cm diameter point that yielded pre-breakdown streamers in a narrow region of potential below breakdown. Shown are a sequence of forty streamers (the blurred brush) that cross the gap showing bright pinpoints of light owing to high fields where they strike the cathode. Note the more luminous secondaries branched and straight within 0.8 cm of the tip. The branching of primaries by the action of space charge in the gap is clearly noted. The branching so attenuates the secondaries that they are visible only about 0.4 cm beyond the axial section. Visually the details are much clearer and the branching and behavior can also be corroborated by a photomultiplier using slits oriented parallel or transverse to the field axis. The return stroke caused by the arrival of a particularly vigorous secondary branch is shown to the left and the branching of secondaries with a sequence of only about four primaries is shown to the right.

Fig. 15 shows one case of branching with only a few streamer sequences superposed and the corresponding spark of Fig. 14 enlarged. The primaries are not visible here. Fig. 16 shows the sequence of events when a 1.9-cm radius sphere and a 1.5-cm gap (that is, nearly uniform field geometry) are used. Here branching does not occur and the streamers and spark are straight and axial. The first sequence crossing the gap leads to the return stroke. The resolution is insufficient to separate the events, but it is clear that primary and secondary cross the gap *from positive sphere* to *negative plane* and that a return stroke follows. Resolution of the primary is so poor that it can only be seen on the last two frames. The lowest trace shows the full amplitude of the return stroke and the time after the primary and secondary for the very bright, arc-like channel fully to develop.

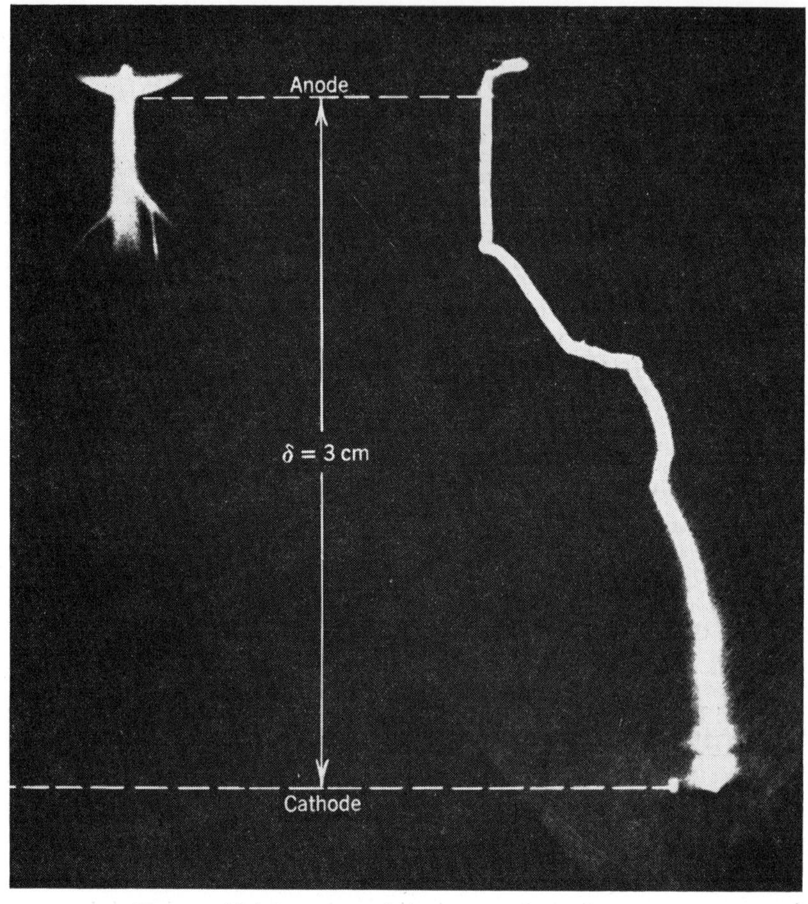

Fig. 15. Enlargements of the two sections of Fig. 14.

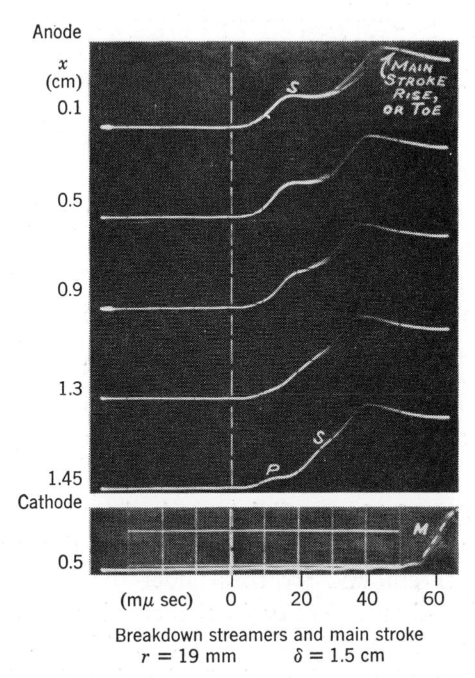

Fig. 16. The events which can be resolved with a large point of 1.9-cm radius and short gap (1.5 cm) with fastest sweep. Since the time resolution is only 7×10^{-9} sec, little can be observed in detail. The fast primary is noted only at $x = 1.45$ cm. The slower secondary appears throughout but is seen to merge with the rise of the light from the return stroke. The cutoff prevents more rise of the main stroke. The lower trace shows this sequence with reduced amplitude, revealing the full height and duration of rise of the main stroke and showing the primary-secondary sequence in proportion at its toe. The sense of motion of the main stroke cannot be determined.

3. POSITIVE STREAMER AND LIGHTNING

The positive streamer mechanism, while not being the dominating mechanism of the lightning discharge stroke, plays a most important role in the lightning discharge mechanism, in that it accounts for the process by which the highly dispersed charge on the raindrops of the negative cloud cell is collected and drains into the channel of the first stepped, or the dart, leaders of the successive strokes and the return strokes.

The negative charge cell of most commonly observed thunderstorms [9] resides in a volume which may have a base as large as 6 km in diameter, but is more often of the diameter of 1 km and extends roughly from the 3-km altitude level above the earth for some 6 km to the 9-km level. The charge in this cell lies on the water droplets totalling from 1 to 10 g of water per m^3 with an average of about 4 g per m^3 in such clouds. The number of drops ranges from 250 to 4×10^3 per m^3. These drops thus range from 0.4 mm to some 1.6 mm in radius. According to Gunn,[9] the droplets of various sizes carry from 0.03 to 0.15 esu of charge, the larger droplets having the heavier charges. Surface electrical fields of the drops range from 300 to 3000 volts per cm, and are thus under the value causing surface instability. The charge density in the cloud ranges from about 40 esu per m^3 to 120 esu per m^3—more or less. The upper figure would yield about 240 coulombs for 6 km^3 of cell with 1 km^2 base and 6 km height. The equivalent positive charge resides on the top of the cloud up to 17 km and somewhat down on the sides.

Studies of Schonland and Malan [1] have shown that such a cell unloads its charge in a succession of strokes to ground, which drain away in succession at each time, the charge remaining in the lowest km of cloud. Actually, it usually drains somewhat less than 1 km in height per stroke, but this figure yields a convenient sample. The time required to discharge such a cloud section is about 0.03 sec. Thus the cell indicated collects the negative charge from the 6-km high cell in six successive strokes taking on the order of 0.03 sec each and delivering about 40 coulombs per stroke to ground.

No one has made any attempt to explain the mechanism by which this can happen until the writer, in 1953,[10] suggested the process which is to be elaborated, but did not then develop it. The problem is to understand how the charge distributed as indicated can be drained into a channel from 20 to 900 cm diameter from 1 km^3 of cloud in 0.03 sec.

In an exceedingly complete and significant study, W. A. Macky,[11] in 1931, showed that when water drops of the same order of size as raindrops find themselves in an electrical field of from 7000 to 10,000 volts per cm, far less than the 30,000 volts per cm required to cause sparks in uniform fields with droplet-free air, the drops become elongated into spindle-shaped figures with intense and long drawn out luminosity in the direction of the negative electrode (cathode) and less of a display at the end towards the positive electrode. These spindles in appropriate fields and drop size may extend some 6 cm towards the cathode. In so doing, they spread over the cathode surface, reaching a diameter of some 3 to 4 cm in the 6 cm of advance. This spreading is accompanied with some very fine water spray, but primarily represents some form of electrical discharge associated with the fine point and does not seriously disrupt the drop. W. N. English,[11] in the writer's laboratory, using hemispherical positively and negatively electrified droplets extruded from capillary tubes of the same dimensions, subjected the resulting filamentary discharges to oscilloscopic analysis. This study revealed that the violent discharge with the flaring filaments from positive points was a manifestation of the common positive streamer actions which have been described. Currents from the droplets ranged from 1 microampere up to 20 microamperes, depending on fields and drop size according to both Macky and English.

This streamer discharge projects positive ions, some positively charged spray droplets, and positive streamer space charge towards the negative electrode in a laterally expanding pattern. These streamers are known to propagate at a speed of 10^8 cm per sec. The luminosity observed by Macky and English represents a succession of such streamers.

In projecting positive charges into the negative droplet region by the deformed droplets, these charges neutralize the negative charges on droplets they encounter, both by the space charge in each streamer channel, by the small positively charged spray, and by positive ions which drift at speeds of 2×10^4 cm per sec or more in these fields. The deformed droplets leave behind the equivalent negative charge on the residue of the droplets from which they came or immediately below and near them. Assuming 10^{-5} ampere of positive corona current per droplet, as observed, the charges on negative raindrops towards the negative cloud cell are dissipated in some 10^{-6} sec and shifted to the lower region. Since the streamers spread laterally in moving upward, increasing areas of the cloud cell are tapped feeding negative charge into a narrower channel below.

Schonland [1] and the writer [10] have independently suggested that the strokes to ground initiate when, through violent turbulence in the throat of the thunderstorm, part of the small positively charged scud cloud is suddenly brought in closer proximity to the negative cell base. This will augment the normal thunderstorm cloud field estimated by Gunn, Chapman,[12] and the writer [10] by different means, to be about 2000 to 4000 volts per cm, up to values of the order of 10,000 volts per cm. In such fields, the large raindrops of the scud cloud will project streamers upward towards the negative cell base. This neutralizes the space charge over the region of high field strength, replacing it by a conducting layer with enhanced negative and positive fields above and below. These distorted fields above and below continue to project positive charge upward, negative charge downward and increase the intervening region of conducting gas. Since spread is lateral as well as axial along the field in the upward streamers, the area drained from the negative cloud cell base will expand laterally as well as axially in the field, neutralizing negative charges above and leaving negative charge below over more and more of the cloud cell. The enhanced fields at the lower positive scud cloud side will eventually become so large that, as noted by Macky and English, one or more drops will yield streamers which develop into spark channels. This serves to concentrate field distortion locally still more and funnels the accumulating negative charge into the now developing pilot leader channel. In this fashion, on the lower, or scud cloud end, a field distortion of such magnitude will materialize that the pilot leader is launched on its earthward progress in a fashion the details of which are at present unknown to us.

It now remains to be shown that the drainage of the cloud base to 1 km depth by such mechanism will occur in some 10^{-2} sec. If the field of 10^4 volts per cm resulting from turbulent displacement of scud cloud mass spreads over a depth of, say, 1 m upward, then the negative drops in this depth are neutralized with existing currents in the order of 10^{-6} to 10^{-5} sec. Thus, in this time, the field distortion and drainage have extended 1 m vertically in 10^{-5} sec so that their upward velocity of advance is at the rate of 10^7 cm per sec. Thus in 10^{-2} sec the neutralized depth will be 10^5 cm, or 1 km as observed. The rate of lateral drainage may be $\frac{1}{3}$ to $\frac{1}{2}$ as fast as the axial drainage, so that the first stroke to ground will perhaps tap less of the lower cell base charge than the later strokes. Once the first return stroke has carried its positive charge up to the cloud base, field distortion up there is such that a new drainage down the preionized channel starts, taking the next km^3 of cell base with it.

4. THE RETURN STROKE

With this indication of the application of positive streamers to the solution of one of the outstanding problems of the mechanism of the lightning stroke, it is now proper to dwell briefly upon the nature of the return stroke. The principle of these strokes, about which little thought has, up to the present, been given, is now successfully being studied on a laboratory scale at Berkeley.

In 1953, the writer [10] considered the mechanism of the propagation of a wave of potential readjustment accompanied by intense ionization following the sudden imposition of a steep potential gradient across an ionized section of conducting plasma.

Assume that as a result of junction of stepped leader and positive earth streamer a very strong potential gradient is suddenly produced across the 20-cm diameter conducting core of the stepped leader channel. In this channel, one can plausibly assume 10^{13} electrons and ions per cm^3 as a result of the stepped leader stroke process. The pressure and density in the channel will be assumed for convenience to be $\sim$760 mm at 300° K. The bright lightning stroke channel initially of 20 cm diameter probably has 10^{17} electrons and positive ions per cm^3. Thus the return stroke must, by means of the potential gradient, suddenly multiply the 10^{13} electrons by a factor 10^4 and propagate its ionization and luminosity up the channel at 10^{10} cm per sec. Electron multiplication proceeds cumulatively for each of the 10^{13} electrons at the rate $e^{\alpha x}$ for a distance x traversed in the field direction. Thus $e^{\alpha x}$ must equal the multiplication factor 10^4 so that αx is roughly 9.3. The quantity α is the number of new electrons created by one electron per cm path in the field direction. Now it can be estimated from the cross section of ionization of electrons of about 10 electron volts average energy that, with the random velocity of the electrons, each ionization in air requires on the average 2×10^{-10} sec. Thus, to create the avalanche of 10^4 electrons with αx ionizing events will consume $9.3 \times 2 \times 10^{-10} = 1.86 \times 10^{-9}$ sec. Now if the ionizing wave is to advance at a speed of 10^{10} cm per sec, the ionizing acts occurring in 1.86×10^{-9} sec must extend over a high field region of $1.86 \times 10^{-9} \times 10^{10} = 18.6$ cm. That is, the ionizing field gradient causing the wave must extend over this distance to give the speed of advance observed, since the ionization time is limited. However, each electron avalanche must be accomplished in much less than 18.6 cm to be consistent with electron drift velocities in the field direction and this time scale. To arrive at the avalanche length, the field strength to pressure ratio X/p, which defines the value of α/p and thus of α, and the drift velocity must be chosen so as to give consistent data. Assuming $p = 760$ mm, and $X/p = 90$ volts per cm per mm Hg, $\alpha = 370$, $\alpha x = 370x = 9.3$, which makes $x = 0.025$ cm. At $X/p = 90$, the electron drift velocity in the field direction is about 2.5×10^7 cm per sec in air. Thus the time for the creation of the avalanche would be roughly 10^{-9} sec, which is, within limits of uncertainty in values of α/p and drift velocities, consistent with that assumed from basic atomic data. The average electron energy at this field is of the order of 9 ev instead of the 10 ev assumed. The value of X/p required at once yields the field strength needed across the 18.6 cm, which is $90 \times 760 = 68{,}400$ volts per cm. Thus the potential across the 18.6 cm provided the field is uniform would need be only 128,000 volts, a thoroughly reasonable figure. Actually, since fields at such junction points would not be uniform and, as in the stepped leader channel, gaseous temperatures are already elevated, the density will be decreased since p is constant, so that corrections are in order and fields may need to be materially higher at one end and extend over longer regions. However, it is seen that to achieve the velocities of return strokes for sparks and the slower lightning return strokes ranging from 5×10^8 to 5×10^9 cm per sec the field distortions are remarkably moderate in size.

In order to attempt to study this propagation under more controlled conditions, R. Westberg,[13] in the writer's laboratory, has just succeeded in observing such pulses in a long glow discharge tube in N_2, N_2O_2 mixtures, air, and in A at around 10^{-2} mm pressure. Fig. 17 shows the apparatus which consists of a 155-cm long tube 4 cm in diameter with Al disc electrodes at both ends. The cathode is at -1740 volts and is in parallel with an 8-μf storage condenser. The anode is grounded through a resistor across which an oscilloscope can determine current before breakdown. A small resistor in series with the power supply at the cathode can suddenly be shorted out, thus applying a small sharp step negative potential increase at the cathode. When the -1740 volts are first applied, a glow discharge in the normal region starts at about 1-ma current. Shorting of the resistor throws the discharge far into the abnormal region with 20

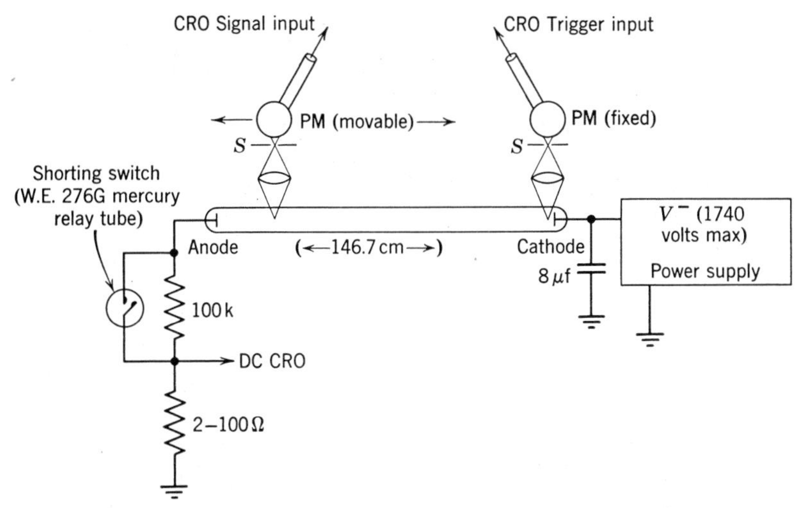

Fig. 17. Westberg's experimental arrangement for studying the motion of luminous and potential pulses down a glow discharge plasma in the transition from abnormal glow discharge of some 30 milliamperes to a transient arc of 200 amperes. The Al disc cathode is at the right end where a power supply and an 8-μf storage condenser are located. The first photomultiplier cell views the bright spot on the cathode that marks the beginning of breakdown and triggers the sweep of the fast oscilloscope. The second photomultiplier observes the sequence of luminous events at various points x down the tube towards the anode at the left. Instead of the photomultiplier, the potential at various points x from the cathode can be observed by the oscilloscope through probes placed at the points, the sweep being still triggered by the initial light flash from the cathode. The shorting switch for overvolting the glow is at the left, as is a slow-sweep oscilloscope to measure tube current during the breakdown process.

to 40 ma current. After 0.1 to 20 seconds' delay through some cathode cleanup action not known, a sudden burst of electrons is liberated from the cathode and observed as a brilliant flash of 0.3-cm diameter on the cathode. This starts a breakdown of the tube to a transient arc in which the current rises to some 200 amp before the condensers discharge and current declines. A photomultiplier cell observes the bright spot on the cathode and starts the sweep on a very fast oscilloscope. A second photomultiplier records the arrival of the luminous potential pulse at various distances from the cathode along the tube. Fig. 18 shows a typical series of oscillograms for different locations. The sweep of time is from left to right. The light causes a downward deflection. At the left are seen the succession of pulses moving from the cathode to the right. In the negative glow and

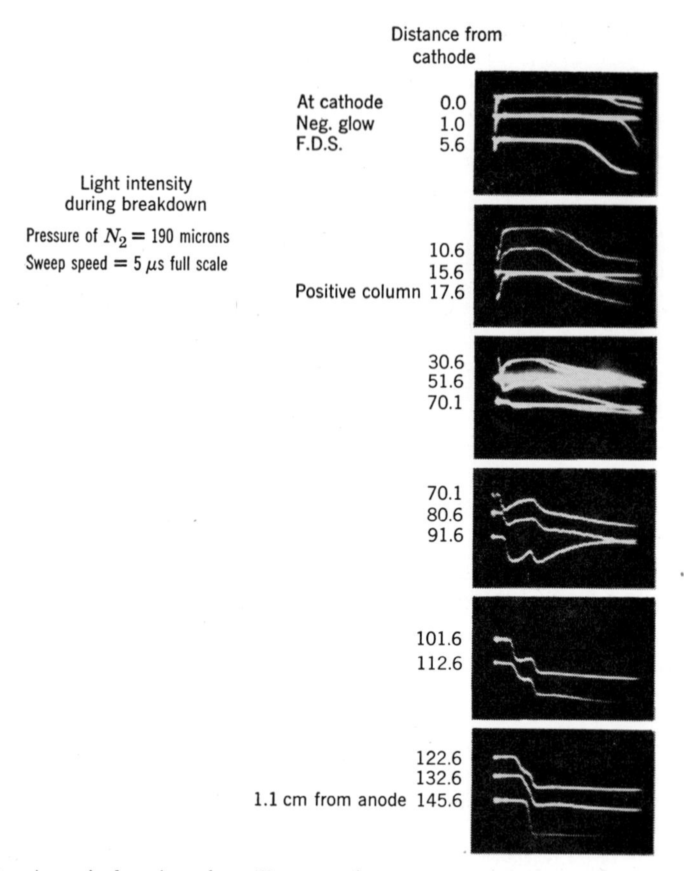

Fig. 18. A typical series of oscillograms for N_2 gas at about 0.19 mm pressure. The sweep in time is from left to right. The points at s cm from the cathode are shown in the vertical series of exposures with the value of s shown at the left. The location of structural features of the glow are also indicated at the left. The sweep speed is 5 μsec full scale. This is too slow to show the progress of a sharp downward light pip starting from the cathode and moving to the right at 2×10^9 cm per sec. In the second frame the slower movement of the pip indicates progress to the right. The changing light intensity requires change in multiplier gain in the successive frames. At 101.6 cm from the cathode the progress to the right is clearly seen. Since the scope is on for a considerable time, events occurring 5 μsec *later* are also recorded. These are the still greater downward *deflection starting from the anode* on the arrival of the first pip from the cathode at the anode. In the top frame it is seen to begin and move from anode to cathode at a speed intermediate between the negative or cathode wave in the Faraday dark space and in the column. The two pips moving in opposite directions appear to meet at s = 145.6 cm or 1.1 cm from the anode. This indicates finite widths of the pulses. Sharp potential drops of the order of 500 volts or more accompany the negative pip moving to the anode. They gradually become less steep as the pulse crosses the gap.

Faraday dark space where electron densities are high, one can hardly discern the movement of the pip from left to right on the slow sweep used. It will be shown at greater speed in Fig. 19. The speed here is 2×10^9 cm per sec. When the wave reaches the positive column with its lower density of electrons, it progresses as can be seen, the speed being of the order of 2×10^8 cm per sec.

Arrived at the anode, it produces a reduction in potential and starts a new quite brilliant wave of luminosity *backward* toward the cath-

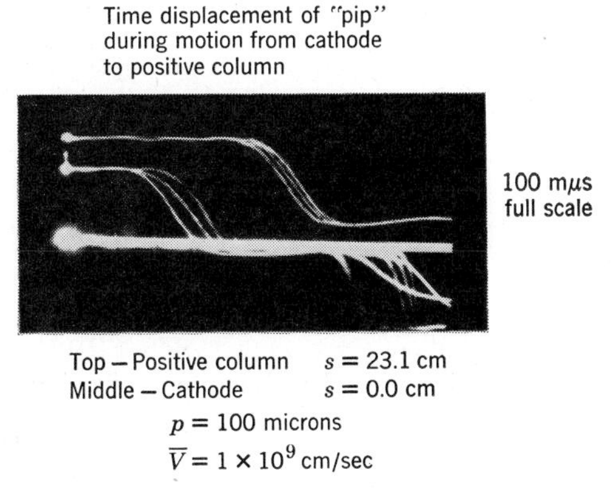

Fig. 19. A fast sweep oscillogram of the pulse triggered at the cathode as observed at the cathode $s = 0$ and at the beginning of the column $s = 23.1$ cm. The three traces show the observed fluctuations for three separate pulses, indicating the inaccuracy of the observations at speeds of 1×10^9 cm per sec.

ode. This is shown in the downward deflection starting from the right. As exposures are taken farther and farther from the cathode, the negative and anode pulses can be seen approaching in time and crossing at 132 cm from the cathode. Since the gradient on reflection at the anode is attenuated, the speed of the return pulse is about 5×10^8 cm per sec. Fig. 19 taken at fast sweep shows the pulses at the cathode and at the end of the dark space, indicating the inaccuracy in time measurements in this region.

Westberg has now succeeded in measuring the potential changes accompanying the luminosity as it advances across the tube by triggering with the first photomultiplier and measuring the potential at various points along the tube by fine wire probes. The potential drop accompanying the brightly luminous pulse is remarkably steep near the cathode where speed is high. It has an amplitude of the order of 500 volts. With the velocity and the temporal change in potential known, the field strength of the potential front as it advances can be measured. The field strength to pressure ratio is ~1000 volts per cm × mm Hg or more and decreases as it advances. It is further possible to observe the redistribution of potential along the whole tube as a function of time as the conducting region sweeps along the plasma at high speeds eventually creating the conditions existing in the transient arc before it extinguishes. The data have now been gathered for air, leaner O_2-N_2 mixtures, very pure A, and will soon be completed in H_2 and O_2. With this information, it is hoped that a quantitative theory of the propagation of such luminosity and conductivity can be achieved. By its means, extrapolation to the higher gas density return strokes of sparks and of lightning discharge channels should be possible. In this fashion, another and perhaps the most startlingly brilliant phases of the lightning stroke will be subject to quantitative description and prediction.

5. CONCLUSIONS

It will be noted that in what has been presented no very detailed picture of the progress of the negative cloud to ground lightning stroke has been portrayed. Although today, we do understand fairly well the mechanism of the streamer type of spark breakdown in air, even to the point of having achieved a quantitative theory for its advance, accurate verification in all its details still remains for the future. On the other hand, you have been presented with newly discovered and hitherto unpublished proof that the common spark is of streamer character and some of its characteristics have been shown. The studies of Allibone and Meek have thrown more light on the nature of the impulsive spark from negative points which partakes more nearly of the character of the lightning stroke. However, today nothing is known in detail about what happens in the sequence of phenomena observed by Schonland with the Boys camera and tentatively interpreted as above. It has been the writer's endeavor to indicate in this paper how the important work of Macky and English and the positive streamer can be used to account for the mechanism of the charge drainage from the cloud in lightning, a hitherto obscure point. It has furthermore been shown that the phenomenon of luminous ionizing potential gradients up the stepped or dart leader channel at speeds approaching one-third the velocity of light can be accomplished by reasonable field distortion and studied in the laboratory with the expectation that an adequate theory can be developed and extended to the return stroke in lightning.

It is by patient and careful study with better techniques as they become available, both on lightning strokes themselves and on their close cousin, the electrical spark, that we hope to push further our understanding of one of nature's most awe-inspiring phenomena, the study of which was initiated by Benjamin Franklin and his kite string some two hundred years ago.

REFERENCES

1. Schonland, B. F. J., and various colleagues, *Proc. Roy. Soc. (London)*, **A 143,** 654 (1934); **A 152,** 595 (1935); **A 162,** 175 (1937); **A 164,** 132 (1938); **A 166,** 56 (1938); **A 191,** 485 (1947); **A 206,** 145 (1951); **A 209,** 158 (1951). *The Flight of Thunderbolts,* Clarendon Press, Oxford (1950).
2. Raether, H., and E. Flegler, *Z. tech. Phys.*, **16,** 435 (1935); also *Z. Physik,* **99,** 635 (1936); **103,** 315 (1936).
3. Raether, H., *Physik. Z.*, **37,** 560 (1936); *Z. Physik,* **108,** 91 (1937).
4. Raether, H., *Ergebnisse der Exakten Naturwissenschaften,* **XXII,** Springer, Berlin (1949), 73–119.
5. Leigh, W., and L. B. Loeb, *Phys. Rev.*, **51,** 199A (1937); Kip, A. F., *Phys. Rev.*, **54,** 139 (1938); **55,** 549 (1939); Loeb, L. B., and J. M. Meek, *Mechanism of the Electric Spark,* Stanford University Press (1941), Chapter II. Loeb, L. B., *Fundamental Processes of Electrical Discharge in Gases,* John Wiley and Sons, New York (1939), pp. 426, 432.
6. Allibone, T. E., and B. F. J. Schonland, *Nature,* **134,** 736 (1934); Allibone, T. E., and J. M. Meek, *Proc. Roy. Soc. (London)*, **A 166,** 97 (1938); **A 169,** 246 (1938).
7. Fisher, L. H., and B. Bederson, *Phys. Rev.*, **81,** 109 (1951); Kachickas, G. A., and L. H. Fisher, *Phys. Rev.*, **88,** 878 (1952); **91,** 775 (1953); Loeb, L. B., *Phys. Rev.*, **81,** 287 (1951); Bandel, H. W., *Phys. Rev.*, **95,** 1117 (1954).
8. Hudson, G. G., Work being completed, to appear in *Phys. Rev.* (late 1957).
9. Loeb, L. B., *Modern Physics for the Engineer,* L. Ridenour (Editor), McGraw-Hill Book Co., New York (1954), Chapter 13; Gunn, R., *Phys. Rev.*, **71,** 181 (1947); *J. Geophys. Research,* **54,** 57 (1949); *Thunderstorm Electricity,* H. Byers (Editor), University of Chicago Press (1954).
10. Loeb, L. B., *Breakdown Streamers and Stepped Leader Strokes,* ONR Report distributed to Special List (June 1953).
11. Macky, W. A., *Proc. Roy. Soc. (London)*, **A 133,** 565 (1931); English, W. N., *Phys. Rev.*, **74,** 179 (1948).
12. Chapman, S., *Phil. Mag.*, **75,** 1333 (1949) and *Cornell Aeronautical Laboratory Report (No. 68)*, (March 1956), pp. 2 ff.; Gunn, R., *J. Appl. Phys.*, **19,** 481 (1948).
13. Westberg, R., Work currently in progress at University of California, paper to appear in *Phys. Rev.* (late 1957).

2 *The Upper Atmosphere*

HARRY WEXLER
Director of Meteorological Research
U. S. Weather Bureau
Washington, D. C.

IV

A Meteorologist Looks at the Upper Atmosphere

FIFTY YEARS AGO, in Philadelphia, the American Philosophical Society celebrated the bicentennial of the birth of Benjamin Franklin. At that gathering Professor Cleveland Abbe of the United States Weather Bureau, referring to Benjamin Franklin as "the earliest of American meteorologists," went on to say:

> It is true that we have records made by observers of the weather before he began his scientific activity, but the progress of meteorology has been such that we have now learned to put the philosophical investigator, that is to say, the man of research, far above the mere observer and recorder. Considered as a mere chronicle of passing events, the study of the weather dates from the earliest ages; but considered as a rational investigation into its ultimate physical causes, or as the logical application of well established principles to the elucidation of unexplained phenomena, or as a system of research checked at every step by observations and experiments, the modern physical meteorology or theoretical meteorology, or dynamic meteorology deals exclusively with force or energy, and dates from the days of Galileo, Sir Isaac Newton, Huyghens, Descartes, Boyle and Gay-Lussac, with whom Benjamin Franklin was a worthy co-laborer. If he had done nothing else but his work in meteorology, that alone would have entitled him to the highest rank. On this subject he thought and wrote for sixty years, from his diary of 1726 to his long-range forecasts of 1786.

Although Benjamin Franklin's atmospheric electricity investigations have attracted most attention, he also made important improvements in meteorological instruments, including the barometer, thermometer, and hygrometer. He studied other important atmospheric phenomena: the thunderstorm, aurora borealis, whirlwinds or dust-devils, waterspouts, and fog formation off Newfoundland. In 1783 he observed that a widespread atmospheric dust "fog" (later found to be of volcanic origin) preceded a very severe winter and used this reasoning as the basis for making long-range weather forecasts.

1. HORIZONTAL AND VERTICAL PROPAGATION OF STORMS

Perhaps of greatest importance to meteorology, in 1743 he made a convincing demonstration of the propagation or progressive movement of a northeast storm along the Atlantic Coast from Philadelphia to Boston, using a lunar eclipse as a time-check. Formerly, it was thought that such coastal storms moved from the northeast to the southwest with the winds. Later, as Postmaster General of the Colonies from 1753 to 1775 he traveled extensively, inspecting reports from his local postmasters about road and weather conditions, the synthesis of which no doubt reinforced his earlier findings about the propagation of storms horizontally along the earth's surface.

Since Franklin's day the horizontal movement of storms has been the main tool which the meteorologist uses in preparing his forecasts. But what can be said of the vertical propagation of storms—either downward from the outer fringes of the atmosphere where solar radiation first encounters appreciable absorption, or upward from the earth's surface where about half of the incoming solar radiation is finally absorbed and transformed into potential, kinetic, and internal energy of the atmosphere? Although some promising clues have been found, the science is still waiting for a Benjamin Franklin to come along with a proof of vertical propagation of storms as convincing as that made over 200 years ago for horizontal propagation.

2. ATMOSPHERIC STRUCTURE AND ENERGY ABSORPTION

Before we consider this question, let us examine some background information about the structure of the atmosphere and the principal layers of energy absorption from solar radiation:

Figure 1 shows the vertical distribution of pressure, density, temperature, and of the principal gases which absorb solar radiation to a

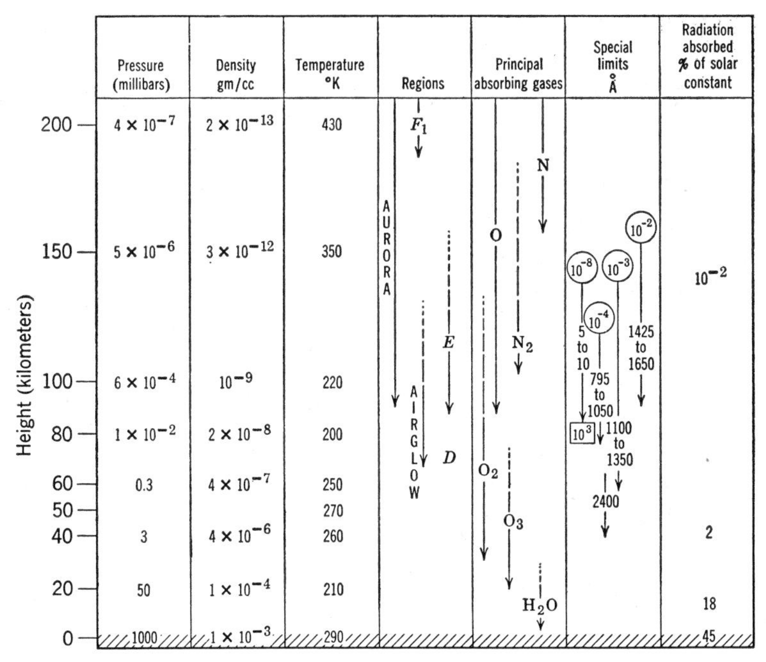

Fig. 1. Principal features of the lower 200 km of the atmosphere.

height of 200 km; it also depicts the spectral limits of the radiation absorbed in the X-ray and ultraviolet regions. These data are taken from a compilation of observations obtained from research rockets (Newell, 1953). The airglow and the *D* and *E* ionized regions are shown as well as the lower portions of the F_1 and aurora regions. The numerals in the circles at the top of spectral limit arrows refer to percentages of the incoming "normal" solar radiation absorbed by the atmosphere in the spectral ranges indicated. The numeral in the box at the bottom of the 5 to 10 A (soft X-ray) range indicates the probable factor of increase in energy of this radiation emanating from a "disturbed" sun. In the last column, it is shown that only 10^{-2}% of the incoming solar radiation is absorbed in the *E* region (100 km), 2% in the ozone layer (20 to 60 km), 18% by water vapor and clouds in the lower 20 km, and 45% by the earth's surface.

The upward decrease of temperature in the troposphere is caused in part by the large absorption of solar radiation at the earth's surface. The warm layer from 20 to 60 km arises from absorption by the ozone of ultraviolet radiation of wavelengths from 1900 to 3100 A. This radiation first undergoes appreciable absorption at the top of the ozone layer; this combined with the low atmospheric density accounts for the relatively high temperature at 50 km.

The general circulation model to 120-km height proposed by Kellogg and Schilling (1951) is shown in Fig. 2. For details of this model the reader is referred to the original paper; the outstanding feature to be noted here is the general convergence and sinking of air over the winter pole. If this sinking were adiabatic, the temper-

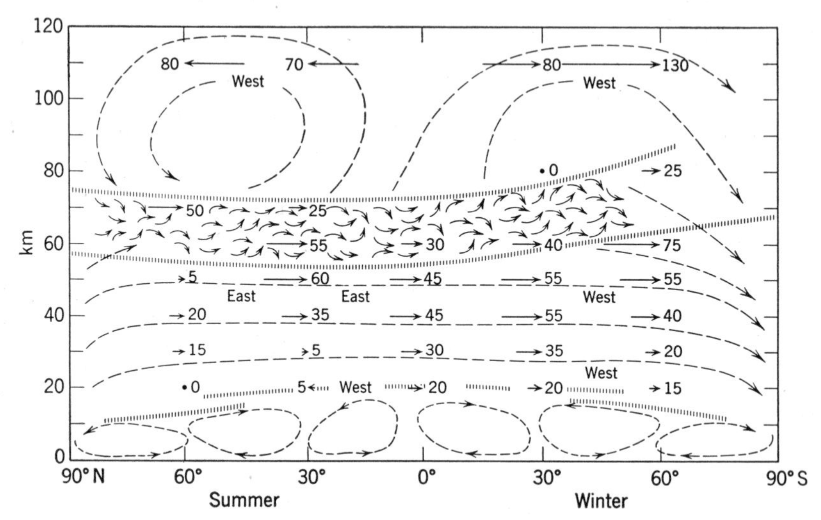

Fig. 2. Schematic representation of general circulation in meridional cross section for northern summer (Kellogg and Schilling, 1951). Numbered arrows refer to zonal winds, speeds in meters per second.

ature of the air would rise appreciably; but in absence of the sun there would be loss of heat by radiation. Whether the resulting temperatures at fixed levels above 30 km would be greater or less than those at corresponding levels at lower latitudes depends on the magnitude of the descending velocities, the temperature lapse rate, and the radiative cooling rate.

Evidence which would appear to favor higher temperatures in polar latitudes above 30 km is given in Fig. 3, which shows a group of early V-2 rocket soundings launched at White Sands, New Mexico (Havens, Koll, LaGow, 1950). The warmest sounding (2), 22 January 1948, occurred during a situation characterized by strong advection of air from the Arctic at all levels up to 12 km, the highest level from which data were available for constructing synoptic weather charts.

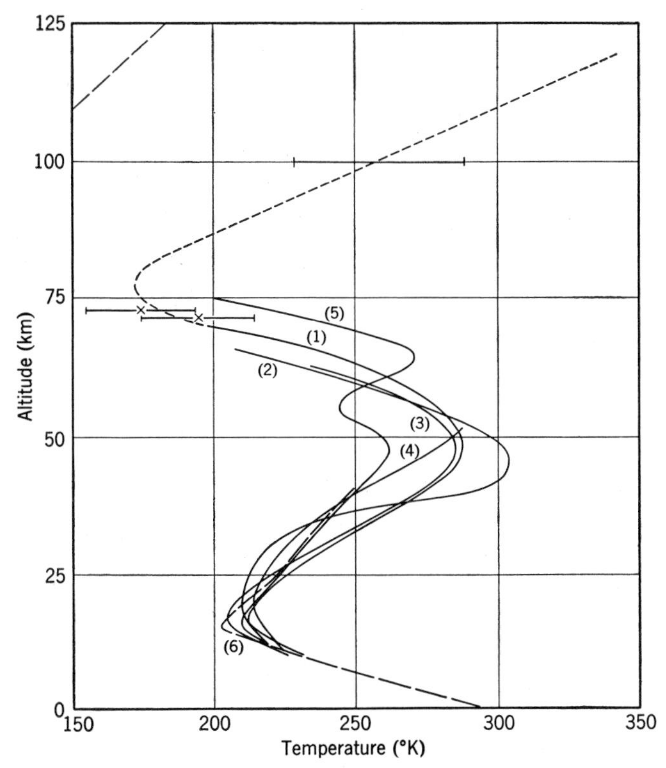

Fig. 3. V-2 rocket soundings, White Sands, New Mexico (Havens, Koll, LaGow, 1950).

(1)	7 March 1947	1123 MST	(4)	28 January 1949	1020
(2)	22 January 1948	1313	(5)	3 May 1949	0914
(3)	5 August 1948	1837	(6)	Radiosonde, Belmar, N. J., 28 September 1948	

The coldest sounding (5), 3 May 1949, was made in a flow of air from lower latitudes, extending at least to 16 km, the highest level of weather chart reliability. If we assume that the lower-level winds are indicative of the flow at upper levels (probably a risky assumption as will be pointed out later), then these two soundings, showing such a large difference in air temperature between 30 and 60 km (maximum difference of 50° C at 40 km), indicate that the warmer air apparently comes from the north and the colder air from the south. This somewhat tenuous evidence would appear to lend support to the one aspect of the Kellogg-Schilling model to which we shall refer later.

3. EMPIRICAL STUDIES OF DOWNWARD PROPAGATION OF ATMOSPHERIC DISTURBANCES

Returning now to the question of propagation downward of disturbances from the outer fringes of the atmosphere, let us consider briefly first those generated in the ionosphere by diurnal solar heating. A good deal of evidence, both from radio probing of the ionosphere and variations in the geomagnetic field at the earth's surface, points to rather large diurnal changes in temperature and wind in the ionosphere, but these changes do not show up as significant disturbances of the sea-level field of wind or pressure. (There are, to be sure, diurnal changes in sea-level winds and pressure, but these are usually so small as not to be detectable without rigorous statistical analysis of large quantities of data. For example, wind speeds caused by the semidiurnal solar tidal motions in the *E* region at 100 km are computed to be 50 meters per second or 200 times greater than those at the earth's surface; the latter are so small they cannot be measured by the conventional anemometer.)

This lack of lower atmospheric response to what must be rather large disturbances in the ionosphere would satisfy most people's intuitive guesses, since the mass of air from the *E* region upward is only one-millionth that of the whole atmosphere. Nevertheless, there have been a good many empirical attempts to correlate ionospheric phenomena with sea-level weather. For example, Martyn and Pulley (1936), Fr. Gherzi (1946, 1950), and the Duells (1948) have published findings about such relationships.

Since the work of the Duells has attracted considerable attention, perhaps a few words are in order: Using the international magnetic index *C* as an indicator of solar corpuscular energy streaming into the ionosphere, the Duells found, for winter, during periods of relative sunspot number less than 40, that such a magnetic disturbance was followed within two to three days by a 2-mb pressure fall in Europe and similar rises in Greenland and Iceland. Craig, who extended this study to most of the Northern Hemisphere (1952), found that this claimed effect reversed itself when looked for in two different groups of eight years each. Furthermore, as will be mentioned later, there is doubt that the international magnetic character index *C* used by the Duells is a true measure of solar corpuscular activity.

Another approach is through sudden ionospheric disturbances (SID) or sudden radio fade-outs caused by enhancement of the *D* layer ionization presumably by enriched solar X-ray or ultraviolet energy. Fritz and Slocum (paper in preparation) correlated SIDs with sea-level pressure profile changes; some of their preliminary results are shown in Fig. 4. Fig. 4 shows for winter and summer the changes in meridional pressure profiles 1, 2, and 3 days after SIDs. The indication of a wave moving poleward and increasing in amplitude, especially in

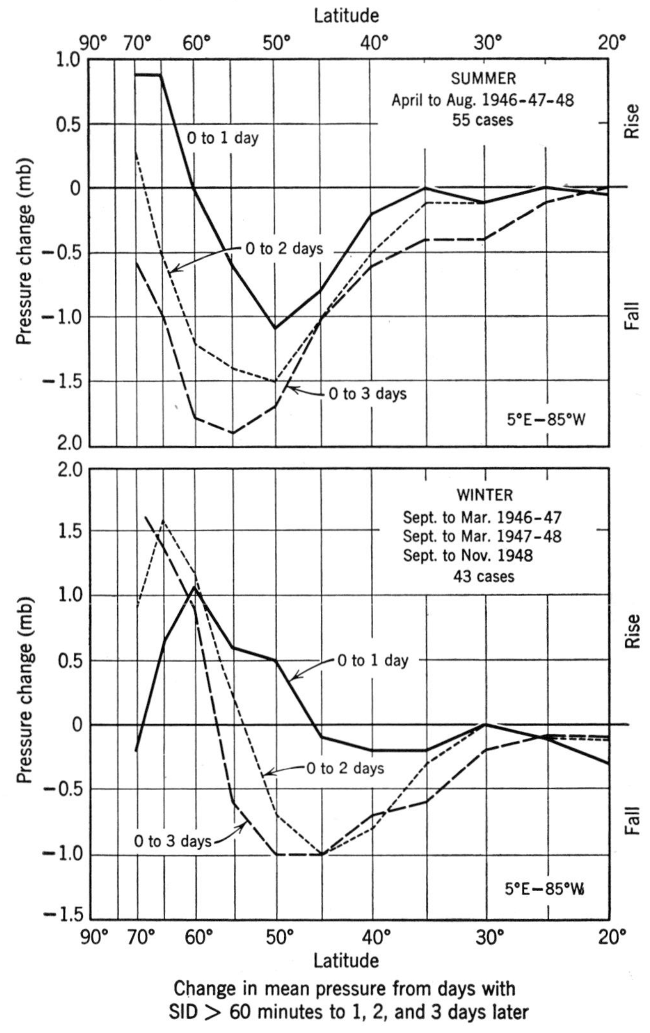

Fig. 4. Average changes in the Northern Hemisphere pressure profile following sudden ionospheric disturbances (SID), winter and summer.

winter, appears impressive at first sight. But a control study for the summers, 1946 to 1948, based on changes in pressure profile three days *before* an SID, showed changes just as large, but of opposite sign, as those occurring three days *after* an SID. One interpretation of this latter result is that the atmosphere knew three days in advance that an SID was to occur and reacted accordingly!

Coming down into the atmosphere to the ozonosphere, above whose base, located near 20 km, there exists 5 per cent of the total mass of the atmosphere, one may feel that disturbances in that more massive layer might have a better chance of propagating downward to the earth's surface. This feeling of confidence is perhaps increased by the fact that the layer of air, whose maximum temperature is found near 50 km, owes its warmth to the absorption of solar ultraviolet radiation by the extremely small amount of ozone which exists in that layer, as mentioned earlier. However, ozone is such a powerful absorber of certain ultraviolet radiations that most of the direct thermal response to changing solar radiation occurs at the top of the ozone layer, near 50 km. For example, from an analysis of the vertical distribution of ultraviolet radiation and ozone observed from rockets F. S. Johnson (1953) found that the diurnal temperature variation has a maximum value of 5.3° C at 48 km and decreases to only 1° C at 31 km and 70 km.

4. THE "BERLIN WARMING"

In recent years Scherhag (1952, 1953), Willett (1952, 1953), and Palmer (1953, 1955) have attempted to ascribe large temperature changes in the layer 10 to 30 km to variations in direct absorption of solar energy—in the form of either ultraviolet or corpuscular radiation. Large temperature changes observed over Berlin, in February 1952 (as much as 45° C in 48 hours at 25 to 30 km), and over Bikati in the equatorial Pacific, in March 1951 (14° C in 48 hours at 10 to 20 km), were alleged to have been caused by enriched streams of radiative or corpuscular energy associated with solar disturbances, as denoted by sudden increases in various solar, ionospheric, and geomagnetic indices.

As an example of the best documented case, that of the "Berlin warming," let us look at Fig. 5, taken from Wiehler (1955). This shows the Berlin soundings taken before and after the marked warming, which occurred in the interval from 0900 February 21 to 0900 February 23. The observations, made by the same trained observers

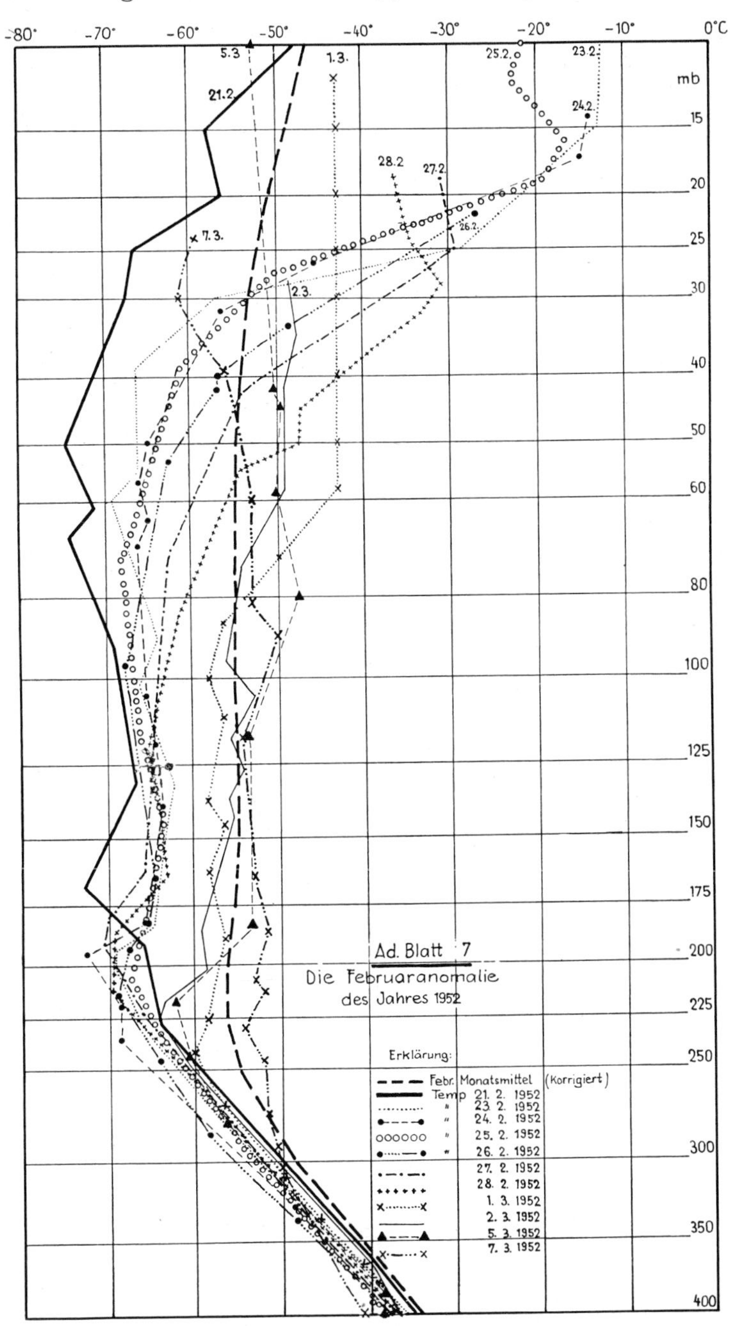

Fig. 5. The Berlin warming of February 1952 (Wiehler, 1955).

using the same type of USAF radiosonde equipment, are considered to be accurate. Because of an unfortunate but understandable coding misinterpretation, some reporters of the event (Schweitzer, 1952; Willett, 1952) stated erroneously that the principal warming occurred from February 24 to February 25, and was thus caused by an immediately previous corpuscular bombardment from the sun. However, as Scherhag later pointed out (1953), the corpuscular radiation responsible for the geomagnetic disturbance originated in solar eruptions at 1400 February 22 arrived at the earth 31 hours after the warming began, and thus could not have caused the warming.

In Fig. 6 are plotted against time the 25-mb temperatures; also the geomagnetic index (K_p), solar flares (SF), and sudden ionospheric disturbances (SID) taken from *Ionospheric Data* published by the National Bureau of Standards, Central Radio Propagation Laboratory, Washington, D. C. The SID (or Moegel-Dellinger Effect) said by Scherhag to have occurred at 1400 on February 22 was not mentioned in the CRPL publication. Note that the principal warming between February 21 and 23 occurred before the magnetic storm began, although some enthusiast might argue that the second warming step followed the MS and CE of the 24th. No persistent warming seemed

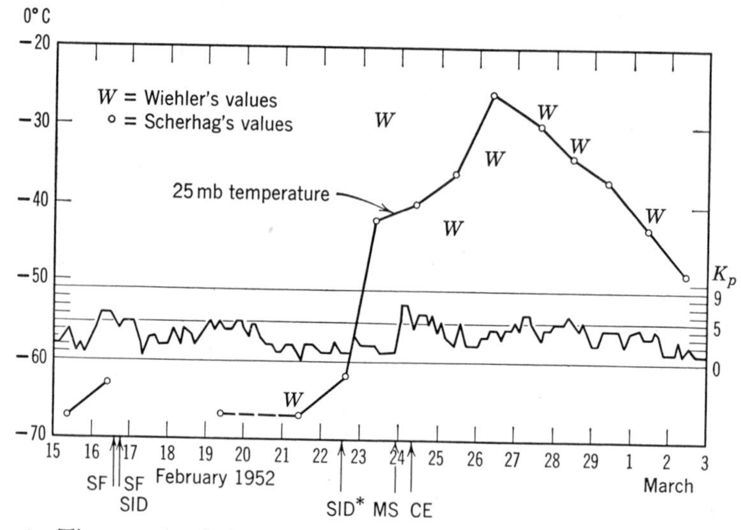

Fig. 6. Time graph of the Berlin warming at 25 mb and daily values of the geomagnetic index (K_p). Also shown are times of solar flares (SF), sudden ionospheric disturbances (SID), chromospheric eruptions (CE), and magnetic storm (MS). SID * signifies an SID referred to by Scherhag but not by the CRPL.

to follow the solar flares and SID of February 16, but the evidence is incomplete because of missing 25-mb temperatures on February 17 and 18.

Scherhag attempted to explain the February 21 to 23 heating by the direct absorption of enriched ultraviolet radiation from the disturbed sun, impinging on very cold stratospheric air of Arctic origin which had a low ozone content; this allowed the radiation to penetrate deeper than normally into the atmosphere before suffering substantial absorption by the ozone, thus causing a large rise in temperature at levels some 20 km below the normal top of the ozone layer.

If Scherhag's explanation is correct, then, depending on thickness of the layer affected by the temperature increase, the observed warming from February 22, 15 hr, to February 23, 09 hr, would require an absorption of energy ranging from $\frac{1}{20}$ ly * min^{-1} to $\frac{1}{6}$ ly min^{-1} if the heating required the *entire* 18 hours. But since Scherhag associates enriched ultraviolet radiation with a solar eruption, one would expect that the major heating would occur within the time span of an SID—say, 1 hour. This would then mean a heating rate ranging from $\frac{1}{2}$ to $1\frac{1}{2}$ solar constants! No one has ever claimed an enriched ultraviolet flux of that intensity.

Ingenious as Scherhag's theory is, however, there appears to be a more logical explanation of the warming based on the ordinary meteorological processes of subsidence of the air to higher pressures leading to adiabatic heating, and accompanied by advection.

In Figs. 7, 8, and 9 are shown contours and isotherms drawn for the 50-mb ($\sim$20-km) surface on February 1, 15, and 29, 1952. At the author's suggestion, daily charts for February and parts of January and March 1952 were constructed by Mr. F. Pooler, Jr., of the U. S. Weather Bureau, and were shown by Dr. Willett in his Atlantic City talk before the American Meteorological Society in March 1953. In spite of the lack of data at 50 mb, the motions and developments appear quite straightforward and regular enough so that the changes in flow and thermal patterns occurring during the entire month can be illustrated by these three charts, two weeks apart.

The first chart, Fig. 7, for February 1, shows a strong cyclone centered over Baffin Bay with temperatures below $-75°$ C and a weak anticyclonic ridge over the Gulf of Alaska with temperatures as high as $-50°$ C. In the course of the next two weeks (Fig. 8) the cyclone moved southeastward over northern Europe, still associated with below $-70°$ C temperatures, while the anticyclonic ridge moved east-

* 1 langley (ly) equals 1 gram calorie per square centimeter.

ward over central Canada, developed into a large anticyclonic eddy, warming as it did so by 10° to 15° C. By February 29 (Fig. 9) the cold cyclone moved into southeast Europe closely followed by the warm anticyclone. The strong horizontal gradient of temperature, produced by the close proximity of the warm anticyclone to the cyclone, moved over Berlin and could thus easily explain the "Berlin warming." According to Wiehler (1955) the 50-mb temperature at Berlin increased by only 8° C from February 21 to 23, while at 15 mb the temperature increase was 45° C! Charts for 15 mb would have been even more interesting than those for 50 mb but it was not possible to draw them because of insufficient data. It took from February 21 to March 3 for the 50-mb temperature at Berlin to increase from −74° to −43° C.

Willett (1953) believed that the warming of the anticyclone over Canada was due to "funneling" or concentration of solar corpuscular energy, in accordance with Menzel's ideas. But these corpuscles can penetrate no lower than the 100-km level and their energy must then be transformed into wave radiation if it is ultimately to make its way deep down into the ozonosphere 60 to 70 km below. No computations have yet been performed to show the magnitude of the resulting heating and the time lag between the two events.

It appears preferable to invoke a more familiar and better understood atmospheric phenomenon, namely, adiabatic heating, associated with subsidence of air layers to higher pressures—a well-known cause of temperature increases at fixed levels in the lower atmosphere. For example, to account for the 40° C rise in temperature observed aloft at Berlin by subsidence alone, an adiabatic sinking of the thermally stable air layers by about 2 km in 48 hours would suffice, or a sinking speed of about 1 cm per sec. Although it cannot be demonstrated that such sinking motion actually occurred in the air which later brought the high temperatures over Berlin, there is support by the following observations of total ozone content at Edmonton, Alberta (Gowan and Leppard, 1953):

TABLE 1. TOTAL OZONE CONTENT, EDMONTON, ALBERTA, 1952
(averaged over periods shown)

Feb. 1–10	0.268 cm	During movement of 50-mb anticyclone into western Canada
Feb. 11–20	.312 cm	Period of maximum warming of anticyclone at 50 mb over central Canada
Feb. 21–29	.299 cm	After anticyclone moved away from Canada

As shown by R. J. Reed (1950), an increase in the total ozone content in a vertical column accompanies a sinking motion, this increase being caused by convergence of ozone aloft into the sinking column of air. For example, as illustrated in Fig. 10, it was computed that a subsidence of the ozone layer between 8 and 16 km at Arosa, with a maximum downward displacement of 1.4 km, would lead to an increase of 0.025 cm in total ozone content, or less than one fourth of the 0.107 cm increase noted at Edmonton from February 10 to 20, during the period of 50-mb anticyclogenesis and presumed sinking of air and its adiabatic warming in central Canada. An explanation of the ozone increase in terms of horizontal advection of air from the east would seem hardly possible because of lack of a source of ozone-rich air at the same latitude.

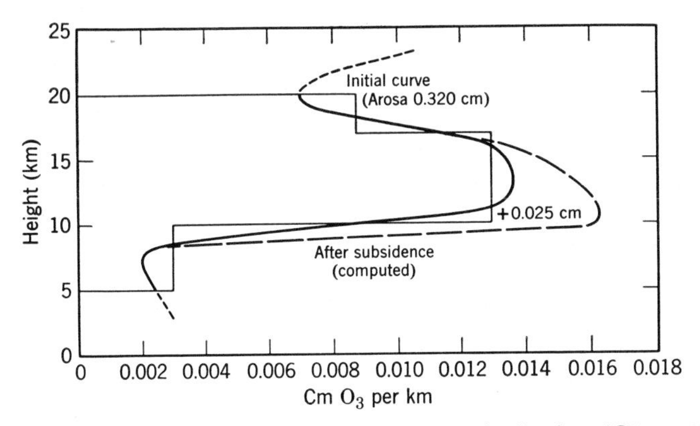

Fig. 10. The effect of subsidence on the ozone distribution (Götz, 1951).

The drop in ozone, 0.081 cm, which occurred at Edmonton from February 20 to 26 is contrary to the well-known seasonal trend, and can be explained by a change in the sign of the vertical motions—from downward during the middle period to upward during the latter period. The 50-mb chart for February 29, shown in Fig. 9, is characteristic of charts during this latter period, which show that the isotherms over central Canada failed to move as rapidly southward from warm to cold as called for by the winds, thus indicating upward motions.

In summary, then, it appears that the Berlin warming in the latter part of February 1952 can best be explained by a combination of subsidence of air beginning over central Canada in mid-February and subsequent motion to central Europe via Greenland hard on the heels of a very cold pocket of air associated with an intense cyclone in the stratosphere.

5. WINDS AND TEMPERATURES IN THE ARCTIC STRATOSPHERE

However, this so-called explanation of the Berlin warming has merely substituted one problem for another. What caused the very cold Arctic stratospheric air (temperatures −75° C or lower) to form in face of the claim by Kellogg and Schilling that on the average the Arctic winter stratosphere should be warm, especially if one believes that the subsiding motions near the winter pole shown in the Kellogg-Schilling model (Fig. 2) are characteristic of daily motions?

Later analysis of daily 50-mb maps for additional winter months shows that at that level there exist at higher latitudes a few large cyclones and anticyclones which move very slowly compared with those at lower levels. These two factors account for far fewer 50-mb pressure centers crossing selected meridians, as illustrated in Table 2.

TABLE 2. NUMBER OF HIGH AND LOW CENTERS CROSSING THE 180°, 90°W, AND 0° MERIDIANS IN A LATITUDE BELT FROM 40°N TO THE POLE, AT THE SURFACE, 500 MB, AND 50 MB FOR JANUARY–MARCH 1953

	180°		90°W		0°	
	High	Low	High	Low	High	Low
50 mb	3	9	1	14	1	10
500 mb	9	36	5	47	14	49
Surface	16	42	29	45	17	41

This striking difference in the number of passages is not caused by lack of data at the top level, since there was fairly good data coverage at the three meridians selected.

The 50-mb maps also showed quite strikingly that the southwest quadrants of these large, intense cyclones are quite cold (−70° to −80° C), much colder than air over the Arctic basin.* Also the isotherms in that quadrant do not move with the wind from day-to-day, indicating that vertical motions exist in this southwest quadrant of the order of 1 cm per sec—first upward as the air moves from the northwest towards the colder isotherms, then downward. On the

* Although the 50-mb charts shown in Figs. 7, 8, and 9 indicate that the cyclones are coincident with the center of low temperature, the more complete charts of 1953 reveal the asymmetry described in this paragraph.

other hand, the anticyclones are much warmer, with temperatures of $-40°$ to $-55°$ C, presumably caused by subsidence. It appears that the temperature distributions within the large cyclones and anticyclones are controlled largely by vertical motions which extend through a considerable thickness of the ozonosphere. If this is so, an independent check could be made by means of ozone measurements, more ozone to be noted in that portion of the southwest quadrant of a 50-mb cyclone where the air is descending and less ozone where the air is ascending. Also, the warming anticyclonic cells should have higher ozone contents, such as was noted in mid-February 1952 over Edmonton, Canada. It is hoped that with the increasing availability of 50-mb charts, this hypothesis can soon be given adequate test on the relatively dense western European network of ozone-observing stations.

From observed vertical displacements of ozone mixing ratio curves, taking into account the time required for establishment of the photochemical equilibrium amounts of ozone, Paetzold found vertical velocities of lower limit of 1 to 10 cm per sec at 30 to 50 km (Paetzold, 1955). Thus it appears reasonable to expect vertical velocities of the same magnitude and their associated adiabatic temperature changes in the layers below.

Even if it should turn out that the thermal fields of cyclones and anticyclones in the 20- to 30-km layer are controlled more by vertical motions than by direct gains or losses of radiant energy or by horizontal advection, we still have the job of explaining these vertical motions. We have fairly reliable dynamic techniques for determining and predicting vertical motions at lower levels where we have more data. In a few years, as our soundings go higher, we shall use these techniques to compute and explain the vertical motions in the 20- to 30-km layer too. The International Geophysical Year of 1957–1958 will provide an excellent opportunity to extend our sounding network into the upper reaches of the ozonosphere, and these soundings, supplemented by an increased ozone network capable of making vertical profile determinations, will help solve these fascinating problems.

6. COMPUTATIONS OF DOWNWARD PROPAGATION OF DISTURBANCES

Returning to the question of the downward propagation of disturbances, let us assume, for the sake of argument, that the temperature of the major portion of the ozonosphere does respond directly and significantly to variations in solar energy; what, then, can be said of the propagation downward to the bottom of the atmosphere of the thermally induced dynamic effects, such as would be indicated by significant changes in the sea-level pressure field?

Charney (1949) showed that in an incompressible atmosphere with a density decrease upward comparable to that of an isothermal atmosphere, a disturbance will propagate downward from top to bottom of the atmosphere at a speed of a few kilometers per day with a strong damping as it does so. Unfortunately, a complete quantitative result has not been worked out for this case, and it is hoped that with aid of an electronic computer, a realistic problem will soon be solved.

In the meantime, starting with an initial thermally induced disturbance, the writer solved a much simpler model (Wexler, 1950). The model was based on the sudden heating of 10° C, and consequent vertical expansion, of a limited portion of an atmosphere symmetric to the subsolar point. Subsequent readjustments of the free and internal surfaces affect the surface pressure profile originally taken as constant. A meteorologically significant maximum pressure drop of 9 mb occurs at the subsolar point, with much smaller rises in the polar wings. However, if all the heating is confined only to the top 5 per cent mass of the atmosphere, corresponding to most of the ozonosphere and all the air above, then, because of the compensating motions of the lower, denser atmosphere, the pressure changes would be reduced by factor of 100 and would thus be too small to be of meteorological significance. But if the heating process is repeated at frequent intervals, such as might occur during bursts of enriched ultraviolet radiation coming from the sun during periods of maximum sunspottedness, and *if* the results are additive, the pressure changes might become meteorologically significant.

The result of a separation of Northern Hemisphere sea-level barometric pressure data into sunspot maximum and minimum periods, for winter and summer half-years, 1899–1939, is shown in Fig. 11. During each of the four solar cycles the pressure at the three consecutive years comprising the sunspot minimum was subtracted from the three consecutive years comprising the sunspot maximum, and the results averaged. In agreement with the model discussed earlier, there are pressure rises in polar regions and falls at lower latitudes. The Northern Hemisphere pressure data are not reliable south of 30° latitude so that we are not able to learn if the pressure fall is largest

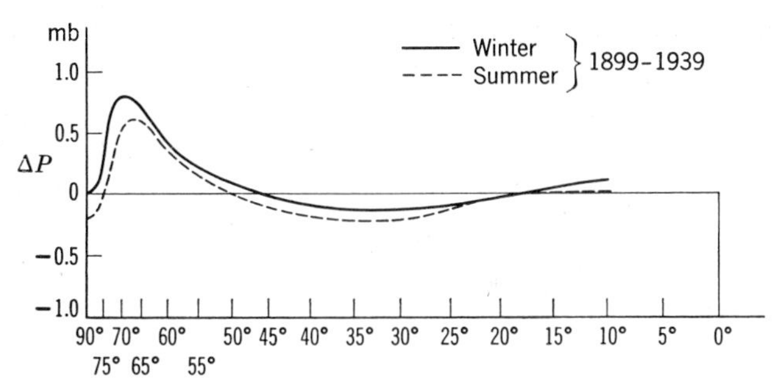

Fig. 11. Mean difference of surface pressure, sunspot maximum minus sunspot minimum, winter and summer, 1899–1939.

near the equator. However, there is a good deal of variation of these pressure-difference profiles from one sunspot cycle to another and this casts doubt on the reality of the effect (Wexler, 1950).

7. UPWARD PROPAGATION OF DISTURBANCES

So much for consideration of the downward propagation of disturbances originating in the upper atmosphere. Now as regards the reverse problem, that of *upward* propagation of disturbances from the surface layer of air to the upper reaches of the atmosphere, we have first the evidence that mountains disturb the atmosphere to great heights. The air flow over the Sierra Nevadas (elevation 4 km), investigated intensively by airplanes and sailplanes, reveals mountain-induced waves in the lower stratosphere at a height four times above that of the mountains. The local adiabatic cooling of the air caused by this lifting lowers the temperature of the lower stratosphere 15° C below the prevailing figure (Kuettner and Rados, 1955).

Even over minor hills of height less than 1000 feet, cirrus cloud formation is noted under proper meteorological conditions, revealing disturbances extending at least to the cirrus level of 10 km (Ludlam, 1952).

That the disturbed air flow over mountains may extend even to the ozonosphere is revealed by the observation of Paetzold and Zschörner (1955) of an "ozone hole" over the Carinthian Alps, when strong winds were blowing from the northeast, perpendicular to the mountains. The total ozone content in a narrow horizontal strip, estimated to be only 20 km wide, amounted to only one-half the value measured in the undisturbed vicinity. This decrease may be due to the effect mentioned earlier, wherein vertical motions, associated with the "mountain-waves," can induce horizontal divergence in the ozone layer, and thus cause a decrease in total ozone content. But to explain an ozone drop as large as 50 per cent without invoking photochemical destruction would require upward displacements of air layers as large as 3 km from their original positions, these displacements taking place in the layer from 10 to 25 km in height. In making these computations Mr. W. B. Moreland, of the United States Weather Bureau, utilized the vertical ozone profiles 1 and 2 shown in Fig. 2 of the paper by Paetzold and Zschörner.

But before one can definitely conclude from ozone measurements that under appropriate meteorological conditions mountain-induced disturbances of air flow can extend to at least 25 km or six times the height of the mountain, more observations are required. It should be emphasized that the measurements of Paetzold and Zschörner were made by observing the sun from one station at various zenith distances ranging from 88° to 79°, and this procedure might introduce uncertainties as to the amount of ozone in the vertical column.

In March, 1956, Mr. T. H. MacDonald of the United States Weather Bureau attempted to find evidence of a similar ozone hole in the region of the famous Bishop wave formed in the lee of the Sierra Nevadas in California. Mr. MacDonald mounted a Dobson ozone spectrophotometer on a truck and made numerous horizontal 11-mile traverses downwind from the mountains during periods of two weak and one moderate mountain waves, in addition to intervening periods of no waves (absence or presence of the waves was deduced by cloud forms and wind velocities). Little or no variation of total ozone content was observed within the range of experimental or observational error. Since the zenith distances in these measurements were always much less than those of Paetzold and Zschörner, the uncertainties raised with regard to the latter's observations would not apply to Mr. MacDonald's observations. Mr. MacDonald's paper will be published in the Proceedings of Symposium on Atmospheric Ozone, Ravensburg, Germany, June, 1956.

If topographical features of horizontal width of only a few hundred miles can extend their influence to the cirrus level and perhaps to the lower ozonosphere, it appears reasonable to expect that large storms can do the same. Air currents, flowing through storms, undergo vertical displacements comparable to those observed near mountains, and should thus disturb a good portion of the ozonosphere. This conclusion is supported by Dobson's well-known relationship of the dis-

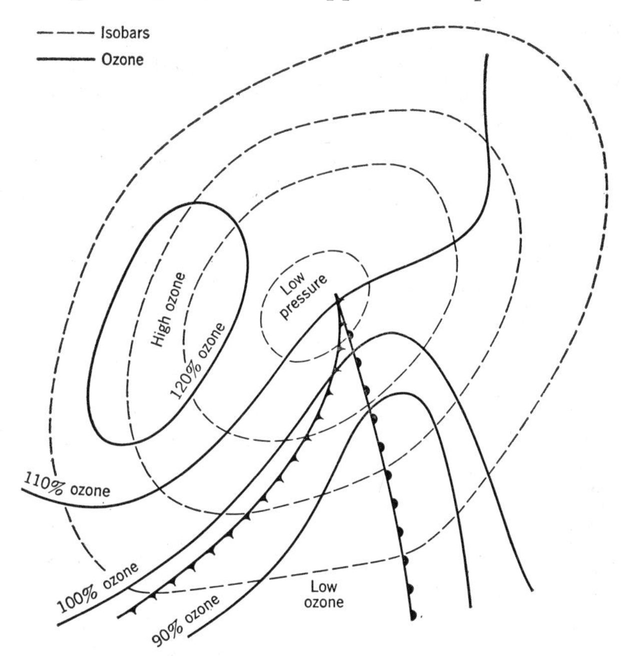

Fig. 12. Ozone distribution about a typical cyclone (Dobson et al., 1946).

tribution of ozone around storms, as illustrated in Fig. 12. Not all of this variation in total ozone content is, of course, due to vertical motions, since horizontal motions of air masses containing varying amounts of ozone also enter. Reed (1950) estimates that the vertical motions account for one third of the total range and horizontal motions for the remaining two thirds.

What can be said of the effect of lower-level disturbances on the ionosphere? That there may be an interrelation here has been suggested by such work as that of Bannon, Higgs, Martyn, and Munro (1940) and Wulf and Hodge (1950). These authors propose a connection between large-scale features of the lower atmospheric circulation and the daily geomagnetic variations, the latter supposedly originating in the lower ionosphere ($\sim$100 km) in accordance with the "dynamo theory." This theory states that the motion of ionized air across the magnetic field of the earth induces electric currents and these, in turn, will cause variations in geomagnetism observed at the ground. Geomagnetic disturbances tend to be stronger near the equinoxes, and Wulf and Hodge have pointed out that it is near these times that maximum instability of the large-scale atmospheric circulation is thought to occur. Large year-to-year changes in the form of the daily geomagnetic variations, averaged by months, also may be associated with year-to-year changes in the general circulation. Now that a series of weather charts of the 50-mb level is available, additional work is required to establish a relationship between circulation patterns and certain geomagnetic variations. This point is of particular importance in establishing cause-and-effect relationships. Some investigators (namely, Duell and Duell, 1948, and Shapiro, 1956) have used *C*, the international magnetic character index, as a measure of solar corpuscular activity affecting the atmospheric circulation; if the ideas of Wulf and of Vestine (1954) are correct, it may turn out that variations in *C* are a result more of changes in the upper atmospheric circulations than of solar corpuscles, and, therefore, correlations of *C* with contemporary or subsequent surface meteorological changes may be the result of dynamic interconnections between lower and upper circulation systems rather than of solar emissions.

We now come to the end of this discussion on the vertical propagation of atmospheric disturbances. There appears to be stronger evidence showing the upward propagation of tropospheric storms to the upper atmosphere than there is for the reverse. The great increase in the number and height of upper air observations planned for the International Geophysical Year of 1957–1958 should make possible a demonstration of the existence and sense of the vertical propagation of atmospheric disturbances as convincing as that made by Benjamin Franklin over 200 years ago for the case of horizontal storm propagation.

REFERENCES

Bannon, J., A. J. Higgs, D. F. Martyn, and G. H. Munro, 1940, The Association of Meteorological Changes with Variations of Ionization in the F_2 Region of the Ionosphere, *Proc. Roy. Soc. (London)*, **A 174**, 298–309.

Charney, J. G., 1949, On a Physical Basis for Numerical Prediction of Large-Scale Motions in the Atmosphere, *J. Meteorol.*, **6** (No. 6), 371–385.

Craig, R. A., 1952, Surface-Pressure Variations Following Geomagnetically Disturbed and Geomagnetically Quiet Days, *J. Meteorol.*, **9** (No. 2), 126–138.

Dobson, G. M. B., A. W. Brewer, and B. M. Cwilong, 1946, Meteorology of the Lower Stratosphere (Bakerian Lecture), *Proc. Roy. Soc. (London)*, **A 185**, 144–175.

Duell, B. and G., 1948, The Behavior of Barometric Pressure During and After Solar Particle Invasions and Solar Ultraviolet Invasions, *Smithsonian Inst. Publ., Misc. Collections*, **110** (No. 8).

Gherzi, E., 1946, Ionospheric Reflections and Weather Forecasting for Eastern China, *Bull. Am. Meteorol. Soc.*, **27** (No. 3), 114–116.

———, 1950, Ionosphere and Weather, *Nature,* **165** (No. 4184), 38.

Götz, F. W. Paul, 1951, Ozone in the Atmosphere, *Compendium of Meteorology,* American Meteorological Society, Boston, Mass.

Gowan, E. H., and R. E. Leppard, 1953, Meteorological Variations in the Quantity of Atmospheric Ozone over Edmonton, *Canadian Journal of Physics,* **31,** 702–713.

Havens, R., R. Koll, and H. LaGow, 1950, Pressures and Temperatures in the Earth's Upper Atmosphere, *Naval Research Laboratory* report, March 1950.

Johnson, F. S., 1953, High Altitude Diurnal Temperature Changes Due to Ozone Absorption, *Bull. Am. Meteorol. Soc.*, **34** (No. 3), 106–110.

Kellogg, W. W., and G. F. Schilling, 1951, A Proposed Model of the Circulation in the Upper Atmosphere, *J. Meteorol.*, **8** (No. 4), 222–230.

Kuettner, J., and R. M. Rados, 1955, "The Structure of the Mountain Wave in the Stratosphere," Talk given at the American Meteorological Society meeting, Honolulu, Hawaii, November 15–17.

Ludlam, F. H., 1952, Orographic Cirrus Clouds, *Quart. J. Roy. Meteorol. Soc.*, **LXXVIII** (No. 338), 554–562.

Martyn, D. F., and O. O. Pulley, 1936, The Temperatures and Constituents of the Upper Atmosphere, *Proc. Roy. Soc. (London)*, **A 154,** 455–486.

Newell, H. E., Jr., 1953, *High Altitude Rocket Research,* Academic Press, New York.

Paetzold, H. K., 1955, New Experimental and Theoretical Investigations on the Atmospheric Ozone Layer, *J. Atm. and Terrest. Phys.*, **7,** 128–140.

———, and H. Zschörner, 1955, Beobachtung eines "Ozonloches" über der Alpen, *Z. Meteorol.*, **9** (H. 8), 250–251.

Palmer, C. E., 1953, The Impulsive Generation of Certain Changes in the Tropospheric Circulation, *J. Meteorol.*, **10,** 1–9; see also various letters in *J. Meteorol.*, **10,** 302–309.

———, 1955, First Synoptic Test of the Solar Hypothesis, Oahu Res. Center, Inst. of Geophysics, UCLA, Scient. Rep. No. 11.

Reed, R. J., 1950, The Role of Vertical Motions in Ozone-Weather Relationships, *J. Meteorol.*, **7** (No. 4), 263–267.

Scherhag, R., 1952, Die Explosionsartigen Stratosphärenerwärmungen des Spätwinters 1951/52, *Berichte d. Deutschen Wetterdienstes in der U. S. Zone,* No. 38, 51–63.

———, 1953, Rätselhafte Temperatursprünge in der Stratosphäre, *Umschau,* **53** (21 Heft), 640–648.

Schweitzer, H., 1952, Haben die Eruptionen auf der Sonne Einfluss auf das Wetter der Erde? *Umschau Wiss. u. Tech.*, **52** (8 Heft), 247.

Shapiro, R., 1956, Further Evidence of a Solar-Weather Effect, *J. Meteorol.*, **13** (No. 4), 335–340.

Vestine, E. H., 1954, Winds in the Upper Atmosphere Deduced from the Dynamo Theory of Geomagnetic Disturbance, *J. Geophys. Research,* **59** (No. 1), 93–128.

Wexler, H., 1950, Possible Effects of Ozonosphere Heating on Sea-level Pressure, *J. Meteorol.*, **7** (No. 6), 370–381.

Wiehler, J., 1955, Die Ergebnisse der Berliner Radiosonden-Hochaufstiege der Jahren 1951–1953, Inst. f. Met. u. Geophys. d. freien Univ. Berlin *Met. Abhand.*, **III** (H. 1).

Willett, H. C., 1952, Atmospheric Reactions to Solar Corpuscular Emissions, *Bull. Am. Meteorol. Soc.*, **33** (No. 6), 255–258.

———, 1953, "Reactions of the General Circulation to Solar Corpuscular Radiation" (paper presented at American Meteorological Society meeting, Atlantic City, March 19, 1953).

Wulf, O. R., and M. W. Hodge, 1950, On the Relation between Variations in the Earth's Magnetic Field and Variation of the Large-Scale Atmospheric Circulation, *J. Geophys. Research*, **55**, 1–20.

HENRY G. BOOKER
Professor of Electrical Engineering
Cornell University
Ithaca, N. Y.

V

Phenomena of Radio Scattering in the Ionosphere

THE PHENOMENON OF REFLECTION of radio waves from the ionosphere has been known for many years and has been studied in considerable detail since the pioneer work of Appleton and Barnett[1] and of Breit and Tuve.[2] It has been shown that there are two main ionospheric layers, one at a height of the order of 100 km known as the E region and the other at a height of around 300 km known as the F region.[3] The total extent of the ionosphere is from a height of perhaps 60 km up to well over 600 km. In this part of the atmosphere free electrons exist, largely as a result of ionization by ultraviolet light from the sun. The electron density reaches a maximum value of the order of 10^4 electrons per cubic centimeter at around 100 km, and the name E region is associated with this maximum. A second maximum occurs at the height around 300 km where the electron density is some ten times the maximum value in the E region. The name F region is associated with this larger maximum. At certain times of day and seasons of the year the F region shows some tendency to separate into two regions known as the F_1 and F_2 regions but with this point we shall not be particularly concerned.

One of the most important ways of investigating the ionosphere is by radar reflections at frequencies from about 20 megacycles per second downwards. By measuring the delay time for these reflections as a function of frequency, ionograms of the type shown in Fig. 1 are made. At the lowest frequencies recorded, the time delay corresponds to reflection from the *E* region. As the frequency is increased, a critical frequency is reached above which reflection takes place from the *F* region. With sufficient increase in frequency even the *F* region is penetrated, and at higher frequencies no reflection from the ionosphere takes place. It will be noticed in Fig. 1 that duplicate traces

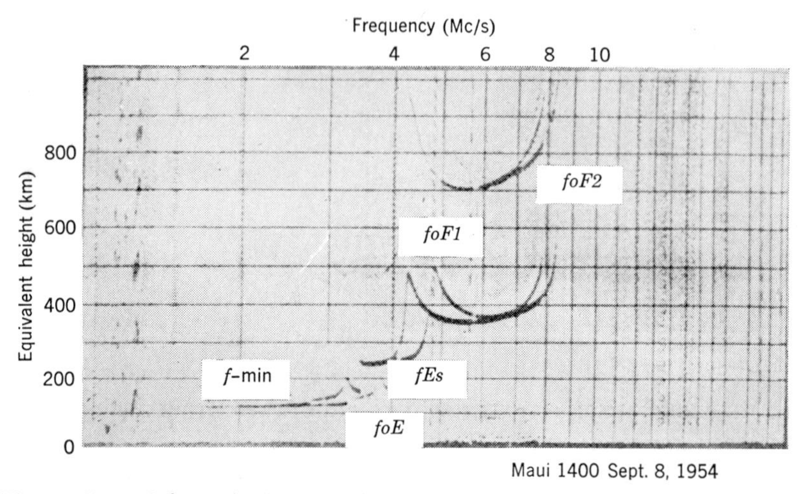

Fig. 1. Record for quiet ionospheric conditions showing echo time (expressed as equivalent height in kilometers) as a function of radio frequency (in megacycles per second).[4]

occur. The duplication parallel to the frequency axis arises because the earth's magnetic field renders the ionosphere doubly refracting for radio waves.[3] The duplication in delay time (or in "equivalent height") arises from multiple reflections between the earth and the ionosphere. Ionograms such as that shown in Fig. 1 are made regularly at many ionospheric observatories distributed throughout the world.

Under quiet ionospheric conditions ionograms show simple clean traces like those in Fig. 1, but there are many situations under which more complicated phenomena occur. I shall describe here the many phenomena that indicate the presence in the ionosphere, either permanently or from time to time, of irregular scattering of radio waves as distinct from classical reflection. Although many of the phenomena to be described have been known for years, in no case is there a generally accepted quantitative explanation available at the present time. In some cases there is no agreement even on the qualitative course of the irregular scattering that is observed. It is quite likely, however, that the phenomena arise from the existence of atmospheric turbulence at ionospheric levels. Turbulence acting in conjunction with some source of ionization could lead to irregularities in electron densities that might explain many of the phenomena. In some cases, however, the phenomena are alleged to occur so high in the ionosphere that it is difficult to see how turbulence could exist at such a great height.

I shall concentrate mainly on describing in outline the various scattering phenomena that have to be explained, although some indication of the explanations now being contemplated will be given in conclusion.

1. FADING OF QUIET IONOSPHERIC ECHOES

Waves reflected from the ionosphere always vary in strength in a manner that is sometimes regular but more frequently irregular. Some of this fading can be explained in terms of interference between the two doubly refracted "magneto-ionic components" associated with the effect of the earth's magnetic field or in terms of interference between a singly and doubly reflected wave. By the use of elliptically polarized antennas and radar technique it is possible, however, to isolate the various modes of transmission and to examine them separately. Let us suppose that in this way we have selected a singly reflected magneto-ionic component for examination. The strength of this wave returned from the ionosphere still fades in an irregular manner over a wide range. Fig. 2 shows the probability distribution of the echo amplitude obtained under quiet ionospheric conditions. Part *a* of the diagram refers to echoes at normal incidence upon the ionosphere and part *b* to reflection at oblique incidence. For normal incidence the range over which the echo fades is not much less than the average amplitude of the echo. Indeed, the probability distribution of the amplitude is roughly of the type known as Rayleigh and is the one expected when scattering from a large number (greater than four) of scattering centers is involved.

This result is quite surprising for the following reason. On the basis of classical reflection from the ionosphere one would not in the first instance expect any fading at all. Allowing for irregular motion due to turbulence or some other cause it is easy to see that some irregular scattering might be superimposed upon the classical reflec-

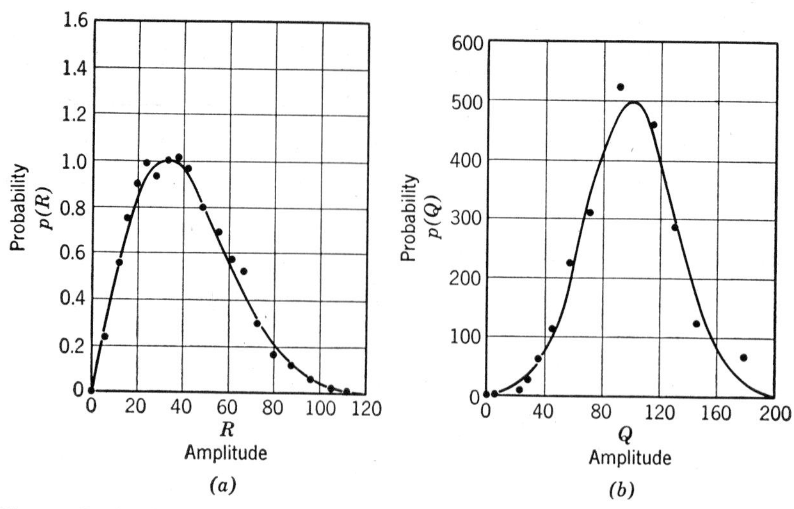

Fig. 2. Probability distribution of echo amplitude under quiet ionospheric conditions: (*a*) normal incidence; (*b*) oblique incidence.[5]

tion but one would have supposed that under quiet ionospheric conditions this would be a small effect. What Fig. 2*a* seems to indicate is that irregular scattering makes a contribution to the radar echo that is nearly as important as classical reflection even under quiet ionospheric conditions.

2. THE "SPREAD *F*" PHENOMENON

Under certain circumstances the trace on an ionogram corresponding to reflection from the *F* region becomes unusually broadened. This phenomenon, which occurs mainly at night, takes a somewhat different form near the equator from that near the poles.[6] Fig. 3 shows three ionograms taken under spread *F* conditions. Ionogram *a* is typical of what happens in high latitudes under spread *F* conditions, *b* in middle latitudes, and *c* in equatorial regions. There is some correlation between the occurrence of this phenomenon and magnetic activity. Although this phenomenon has now been observed for a quarter of a century, its explanation is still largely a mystery despite the evidence which implies that irregular scattering must in some way be the cause.

Rough measurements have been made of the extent to which the direction of arrival of the echo departs from the vertical. Under spread *F* conditions this direction wanders from the vertical to a some-

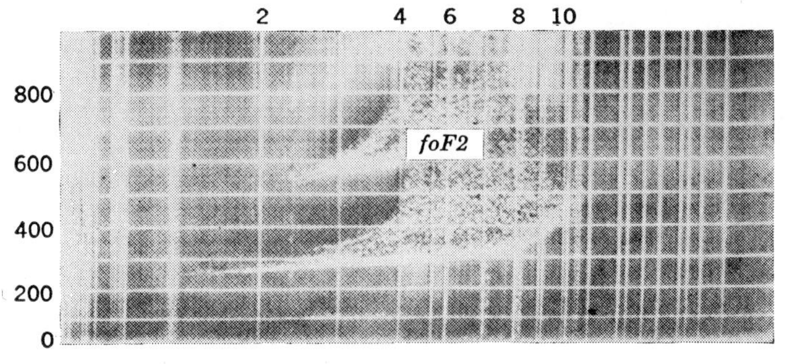

Spread echo of the type commonly observed at high latitudes.
Godhavn 1640 Dec. 22, 1951

(*a*)

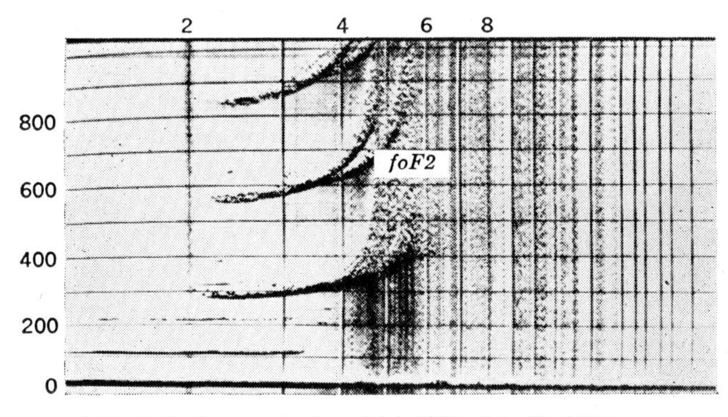

Middle latitude spread echo. Adak 2015 July 10, 1955

(*b*)

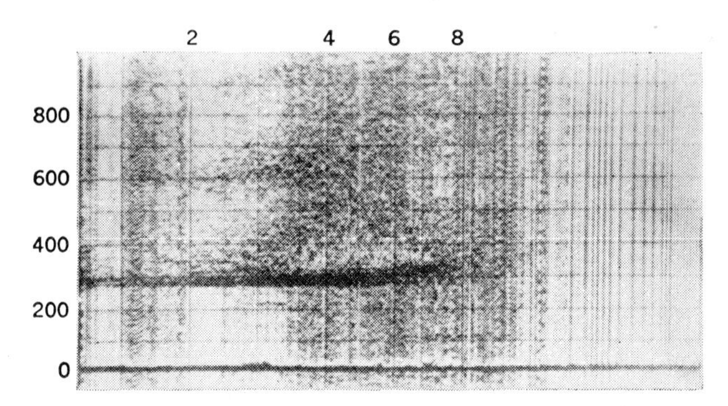

Equatorial-type spread echo. Huancayo 0400 June 3, 1952

(*c*)

Fig. 3. Ionospheric records showing "spread *F*" conditions: (*a*) in high latitudes; (*b*) in middle latitudes; (*c*) in the equatorial regions.[4]

what greater extent than under quiet ionospheric conditions.[7] This effect, however, is too small to explain the spread in range in terms of a variation in slant range to scatterers at a roughly fixed height.[8]

3. ASPECT-SENSITIVE ECHOES FROM THE *E* AND *F* REGIONS

Comparatively recently it has been discovered that echoes can be obtained from the ionosphere from directions approximately perpendicular to the earth's magnetic field. Suppose that perpendiculars are drawn from an ionospheric observatory on to the various lines of the earth's magnetic field. Let us now select the feet of perpendiculars that are at a fixed level, such as 250 km. These feet lie along a curve at the 250-km level which is symmetrical about the magnetic meridian. It has been found that scattered echoes from the vicinity of such a line are frequently obtained at night. The conditions under which this occurs may well be those which produce spread *F* on an ionogram. It has also been found, however, that similar echoes occur at *E*-region levels, and these are illustrated in Fig. 4. These observations suggest that there exist, in the *E* and *F* regions, irregularities of electron density that are elongated along the earth's magnetic field

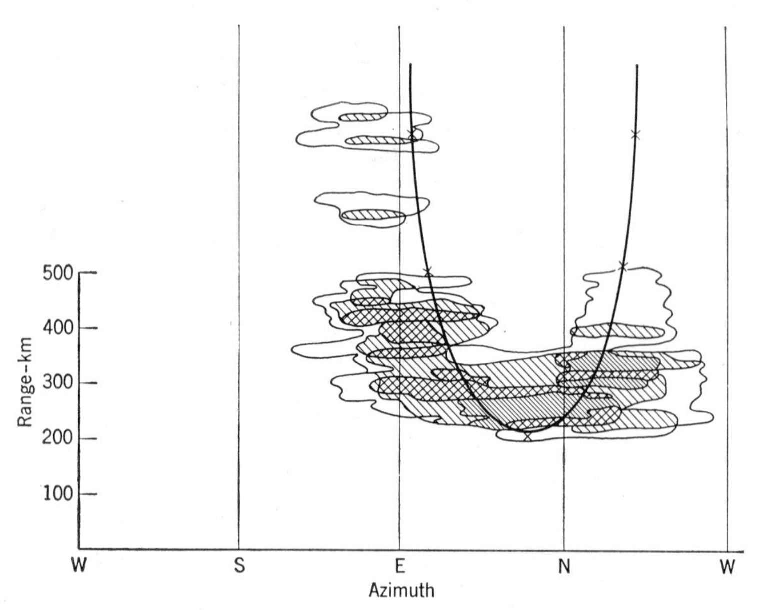

Fig. 4. Variation of range with azimuth for 23 megacycles per second echoes centered on magnetic north.[9]

and which consequently produce an unusually large amount of back-scattering when viewed perpendicular to their length.

4. TWINKLING OF POINT SOURCES OF COSMIC RADIO NOISE

There are several discrete sources of cosmic radio noise, of which one of the most intense is in Cygnus. If the noise from one of these discrete sources is observed in the VHF band, the strength frequently shows irregular variations of the type that would be called twinkling in the case of an optical star.[10, 11] Twinkling occurs principally at night and appears to show some correlation with the occurrence of the spread *F* phenomenon. Fig. 5 shows observations of the twinkling of radio noise from the discrete source in Cygnus on two occasions when Cygnus was setting and on one when it was rising. It will be observed that the twinkling becomes more marked at low angles of elevation. This twinkling of radio stars almost certainly occurs in the ionosphere and it has been shown that the irregularities of electron density required to explain it must be elongated along the earth's magnetic field.[34] The height in the ionosphere at which it occurs is uncertain. Arguments have been advanced for thinking that the main phenomena occur above the level of maximum electron density in the *F* region.[10, 11] Work in Australia,[13] however, has shown that it is possible for twinkling of radio stars to occur in the *E* region under conditions of sporadic ionization (see section 9).

5. LONG-DURATION METEOR ECHOES

A meteor entering the atmosphere produces an ionization trail extending from perhaps 100 km down to perhaps 80 km, and a VHF radar equipment can, under suitable circumstances, obtain an echo from the trail.[14] For most trails it is necessary, in order to obtain an observable echo, for the foot of perpendicular from the radar on to the line of the trail to lie between about 80 and 100 km in height. Even then the echo usually disappears after a few tenths of a second because of diffusion of the trail. In the case of unusually strong meteor trails, however, it is possible for the radio echo to last many seconds or even many tens of seconds. For these long-duration meteor echoes perpendicularity from the radar on to the original trail is not necessary. Fig. 6 shows the variation with time of the echo time from a meteor trail. The parabolic curve is associated with the formation of the trail by the incoming meteor, and the minimum point

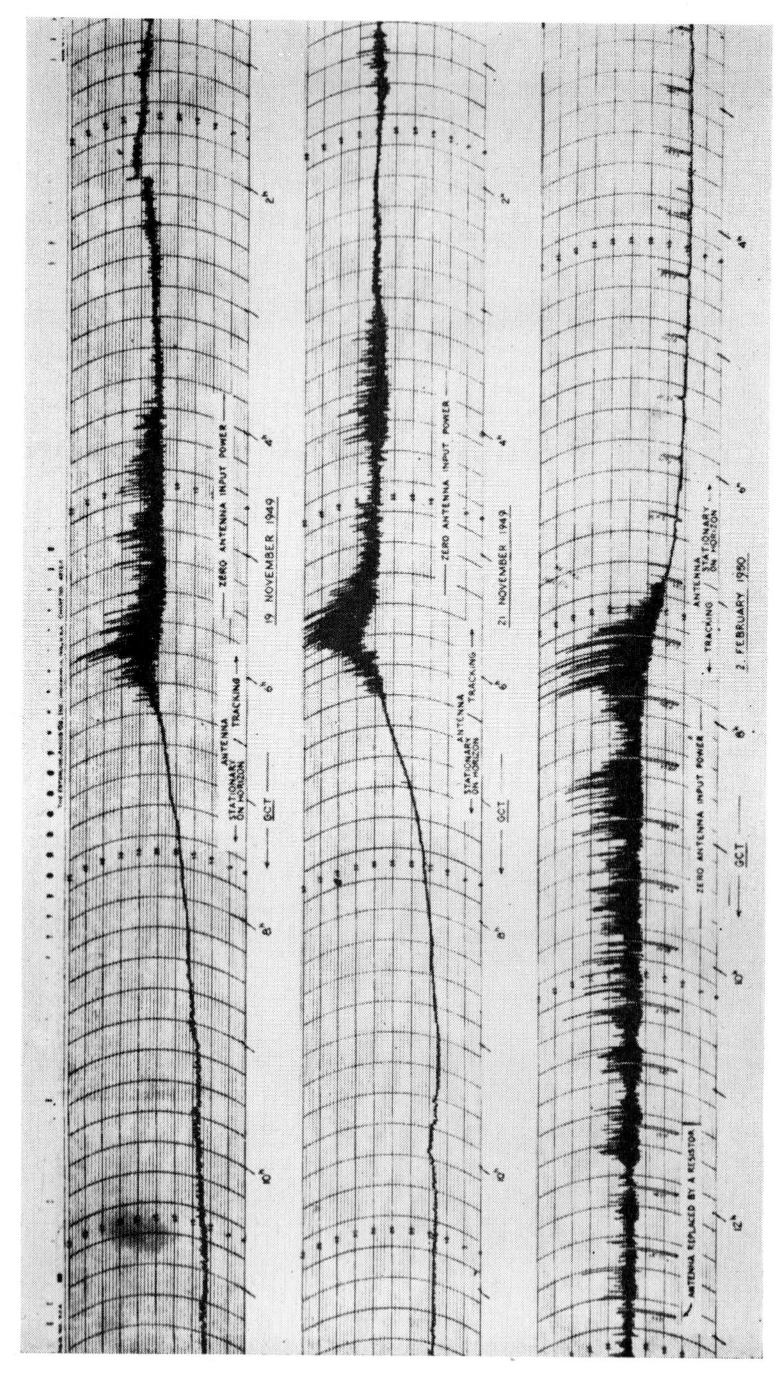

Fig. 5. Twinkling of discrete source of radio noise in Cygnus during rising and setting.[12]

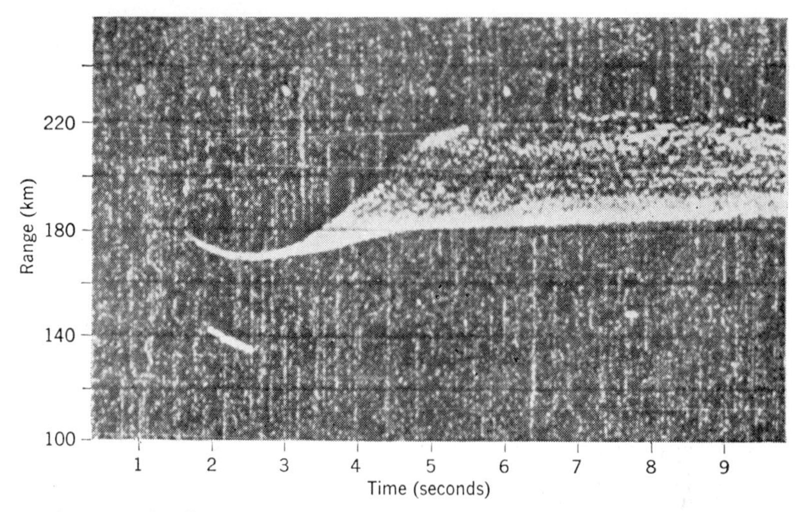

Fig. 6. Echo from meteor trail showing variation of range with time.[35]

in this curve indicates the time at which the meteor passes the foot of the perpendicular from the radar on to the line of the trail. It will be observed, however, that an echo exists for some appreciable time after the passage of the meteor. During the first tenth of a second or so after fall of the meteor the echo frequently shows certain special fading phenomena such as those indicated in Fig. 7: these are caused by the diffraction phenomena associated with formation of the trail and disappear when the trail has been completely formed. However, if the echo lasts more than a few tenths of a second, deep fading of a more irregular nature develops as shown in Fig. 8.

There is little doubt that the fading of long-duration meteor echoes is associated with atmospheric turbulence at the levels where meteor trails occur. Winds observed at these levels by observation of radar echoes from meteor trails show great variability as shown in Fig. 9. What is measured in Fig. 9 is the instantaneous wind at a particular

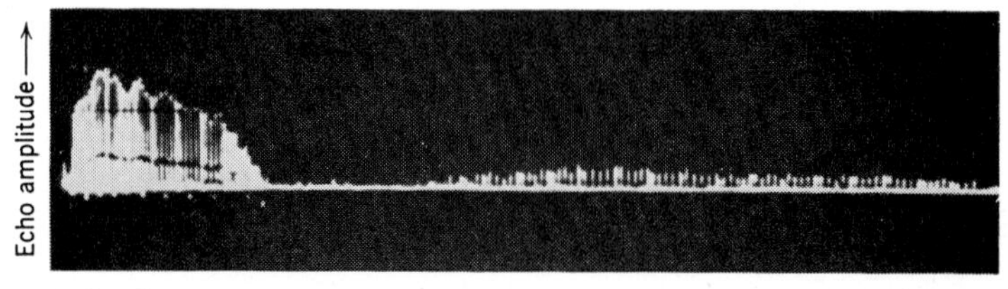

Fig. 7. Echo from meteor trail showing variation of amplitude with time during first two tenths of a second (750 pulses per second).[14]

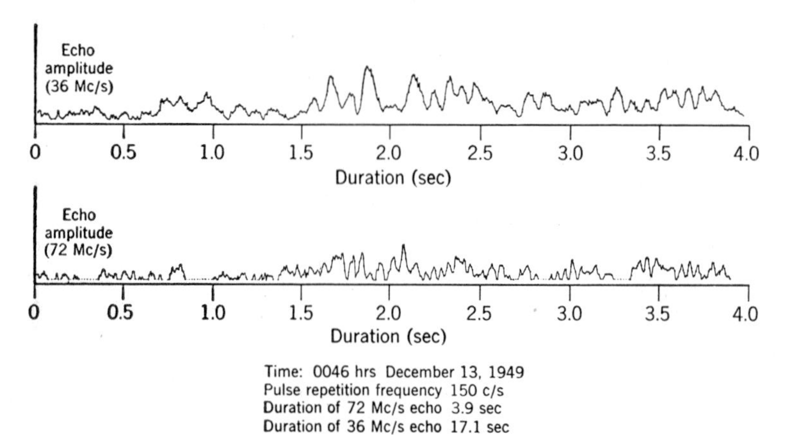

Fig. 8. Fading of long-duration echo from a meteor trail observed simultaneously on two radio frequencies.[15]

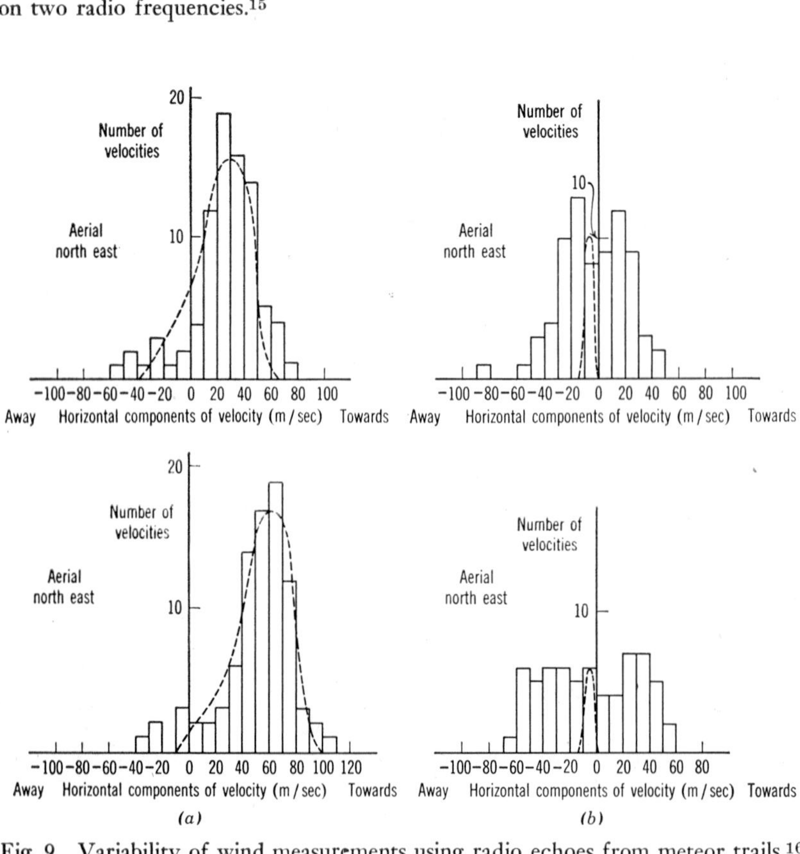

Fig. 9. Variability of wind measurements using radio echoes from meteor trails.[16]

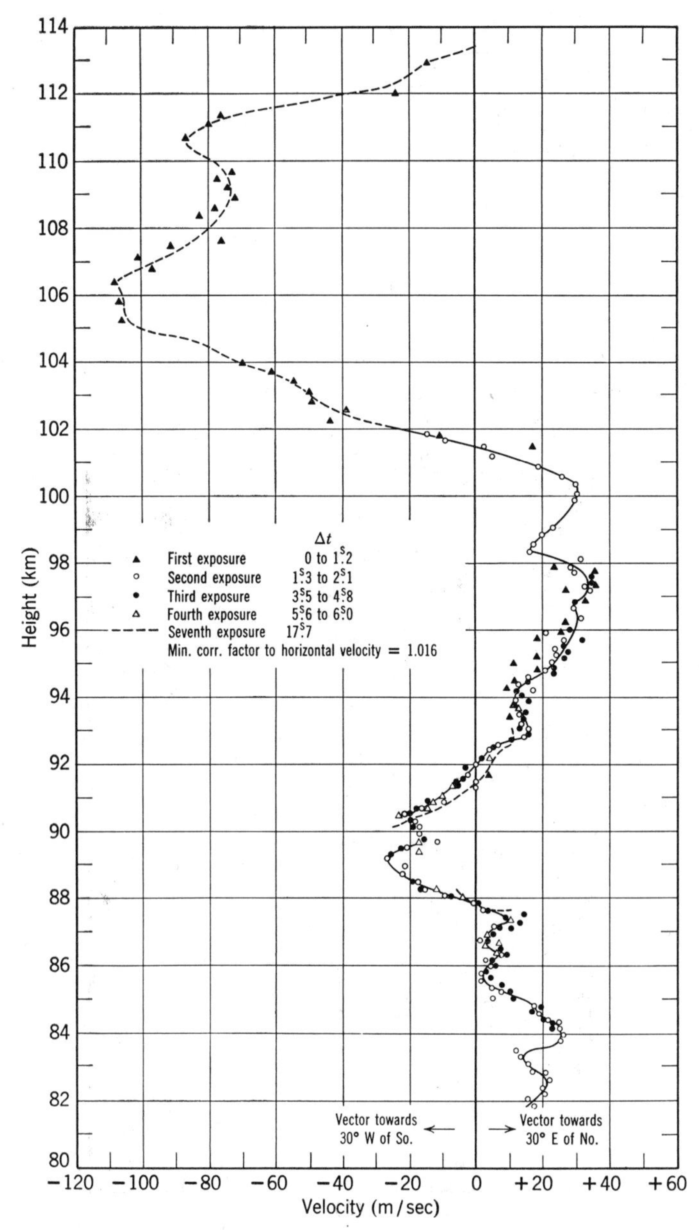

Fig. 10. Variability of instantaneous wind measurements with height using photography of meteor trails.[17]

instant, that is, the vector sum of mean wind and turbulent motion. The variability in the wind shown in Fig. 9 is therefore an indication of turbulence velocity and shows that turbulence velocity at these levels is not much less than the mean wind speed. Furthermore, photographic observations of visible meteor trails by Whipple have demonstrated the irregular nature of the motion at different points along a meteor trail (see Fig. 10). Finally, a number of photographs have been made, of which that shown in Fig. 11 is one, showing visible meteor trails of long duration, and these pictures show the

(*a*)

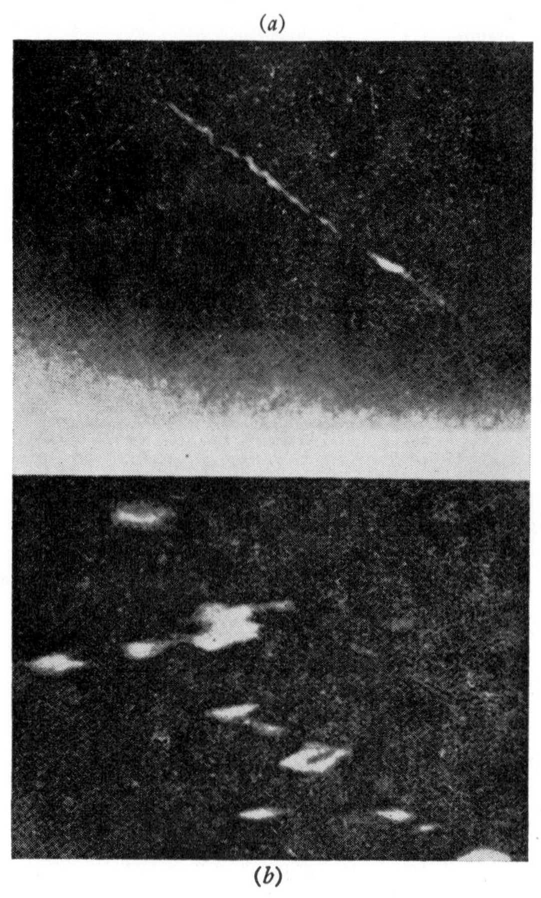

(*b*)

Fig. 11. Photographs of long-duration visible meteor trail: (*a*) after a "short" time (probably a few minutes) from fall of the meteor; (*b*) after about half an hour.[18]

effect of eddy diffusion upon the trail in a manner somewhat similar to that observed with smoke in the lower atmosphere.

6. ECHOES FROM BELOW THE *E* REGION

With normal ionospheric recorders persistent echoes are not usually obtained from levels below the *E* region (100 to 120 km in height). With more sensitive equipment, however, echoes have been obtained at frequencies of the order of 2 to 5 megacycles per second from levels below that of the *E* region. Fig. 12 shows an example of echoes of

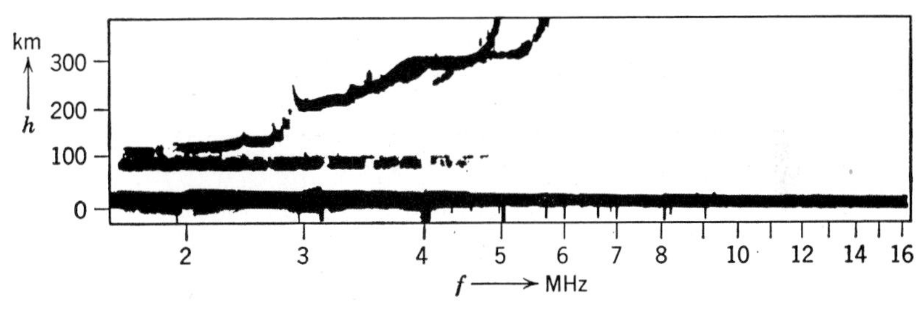

Fig. 12. Ionospheric record showing echoes from an equivalent height less than 100 km.[19]

this type from a level of around 80 km. With extremely sensitive equipment operated in Australia under conditions of very low noise level, echoes have been obtained from more or less all heights extending from the *E* region down to a height between 60 and 70 km. This is illustrated in Fig. 13, which shows a typical example of the variation of echo strength (on a logarithmic scale) with height. Echoes from

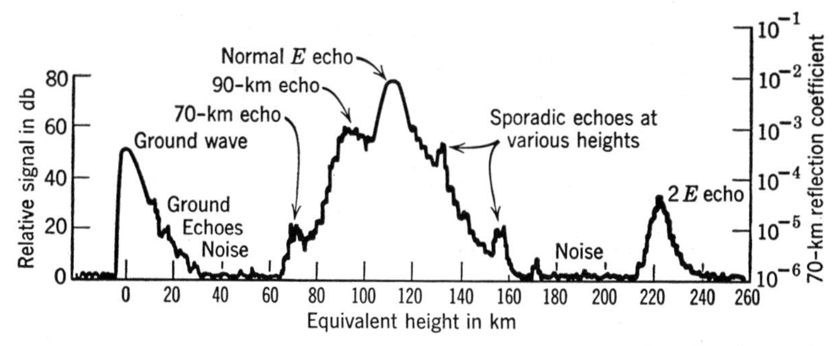

Fig. 13. Variation of echo amplitude with equivalent height, showing echoes from an equivalent height less than 100 km.[20]

below the *E* region fade at an appreciably faster rate than regular *E* region echoes as illustrated in Fig. 14. The echoes from heights between 70 and 100 km are continually fading between a maximum value and the noise level, and there is little correlation between the fading from two levels separated by, say, 10 km. Echoes are particularly strong in the neighborhood of 90 km at night with a somewhat lower height prevailing during the daytime.

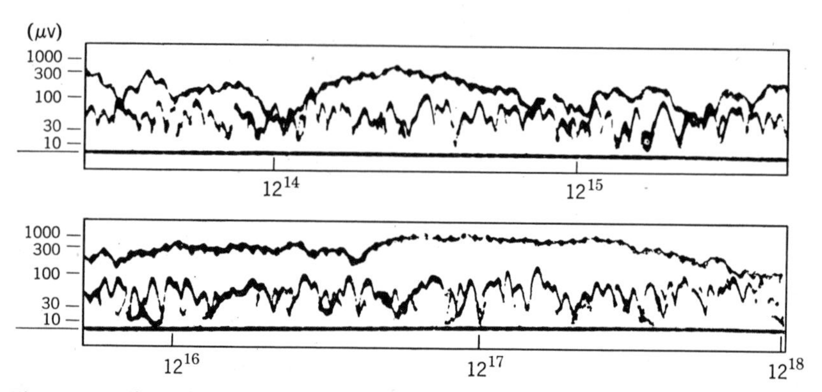

Fig. 14. Fading of echo from the regular *E* region (upper trace) compared with fading of echoes from below the *E* region (lower trace).[19]

7. VHF SCATTER COMMUNICATION

At frequencies of the order of 25 to 100 megacycles per second it has been found that, with antennas that give a sufficiently narrow beam, communication can be maintained over distances of the order of 1000 to 2000 km. If one assumes classical reflection from the ionosphere, such communication should normally be impossible. The signal actually received is weak but reliable. The fading shows a Rayleigh distribution of amplitude and the phenomenon almost certainly arises from scattering in the ionosphere. Measurements of the height from which this scattering appears to take place indicate a height of the order of 90 km at night with a somewhat lower value prevailing during the day.[22] A record of the signal strength at 50 megacycles per second over a 1200-km path is shown in Fig. 15.

The spikes are due to specular reflection from meteor trails before turbulence makes them irregular. The cause of the background signal is, however, undecided at the present time. It may be due to specular reflection by a large number of weak meteor trails that have not been

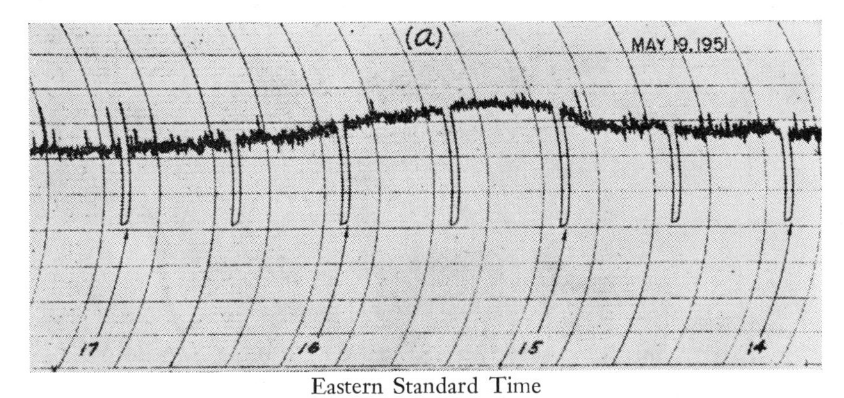

Eastern Standard Time

Fig. 15. Signal received at 50 megacycles per second over 1200-km path.[21]

rendered irregular by turbulence.[23] It is more likely, however, that the background signal arises from scattering by irregularities of electron density caused by turbulence.[24] The ionization that is rendered irregular by turbulence could, however, be of meteoric origin. The diurnal variation of the signals received are such as to suggest that the ionization that is rendered irregular by turbulence is of meteoric origin at night but is primarily of solar origin during the day.[21]

8. ECHOES FROM AURORAL IONIZATION

During auroral activity VHF scatter communication is usually enhanced as shown in Fig. 16. At such times the fading rate of the signal is increased by a power of ten or more, and the direction of arrival is shifted from the great circle path in such a way as to suggest that the predominant scattering is then taking place somewhere in the vicinity of the visual auroral activity. Scattering from ionization apparently associated with auroral activity can also be seen, under certain circumstances, by VHF radar equipments whose beams are directed towards the auroral activity. Fig. 17 shows three photographs of the plan position indicator of a 100-megacycles-per-second radar receiving such echoes. On these plan position indicators range is plotted radially on the scale shown and the northerly direction has been arranged to be downward. Under each radar picture a photograph of the visual activity in the northern sky is shown. Since angle of elevation in these pictures corresponds to distance toward the north, a range and azimuth scale can be marked on them approximately as shown. By comparing the radar and visual pictures, a comparison can

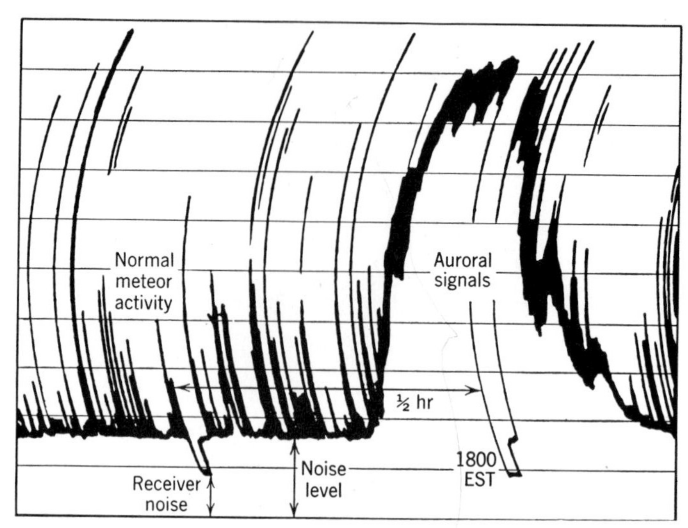

Fig. 16. Signal received at 50 megacycles per second over 1300-km path under auroral conditions.[25]

be made between the locations from which auroral light is coming and the locations from which radar echoes are coming. In this way it is found that there is significant, but not complete, correlation between the visual and radar phenomena.[26]

An important feature of radar phenomena is that echoes are only obtained at low angles of elevation in roughly the northern quadrant even though there may be visual auroral activity more or less all over the sky.[27] This is illustrated in Figs. 18 and 19, which refer to a 50-megacycle-per-second radar used at the Geophysical Institute of Alaska. In Fig. 18 it is seen that the locations of the echoes are in a 120-degree sector centered on magnetic north. In Fig. 19 it is seen that most auroral echoes have ranges in the vicinity of 400 to 1000 km. A cutoff at long range is expected for normal horizon reasons. The comparative absence of short range echoes in the vicinity of 100 to 400 km is, however, a significant feature of auroral radar echoes in the VHF band. Similar results obtained at Point Barrow are shown in Fig. 20.

Another striking feature of auroral echoes is the comparative rapidity with which they fade. Fig. 21 illustrates the fact that auroral echoes on 50 megacycles per second fade at a rate of the order of 200 cycles per second. If this is interpreted as due to Doppler shifts arising from some kind of motion in the ionosphere, one would deduce

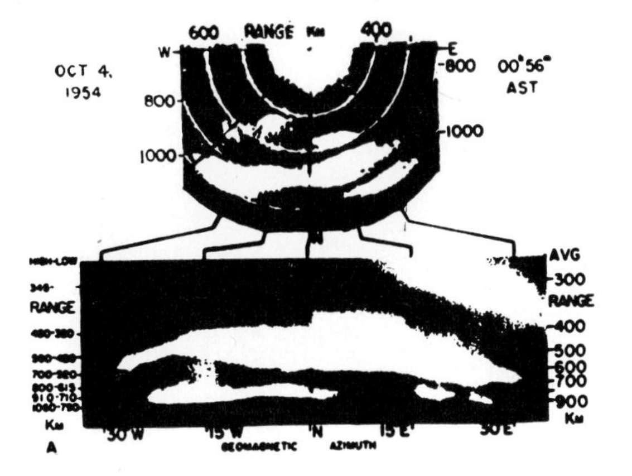

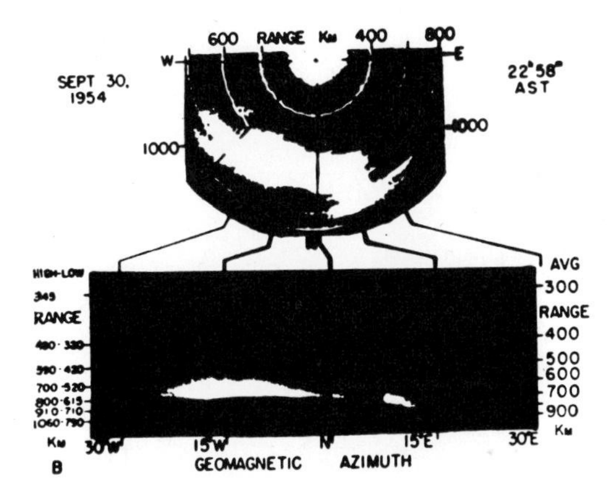

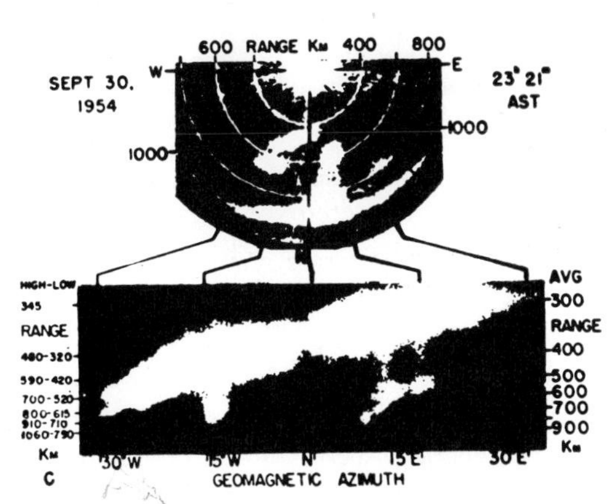

Fig. 17. Simultaneous pictures of radar echoes from aurora and of visual auroral activity.[26]

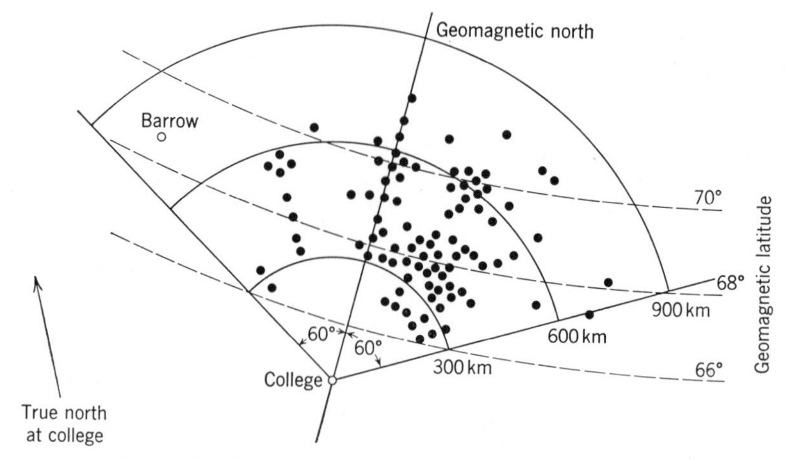

Fig. 18. Azimuth distribution of auroral echoes (50 megacycles per second).[28]

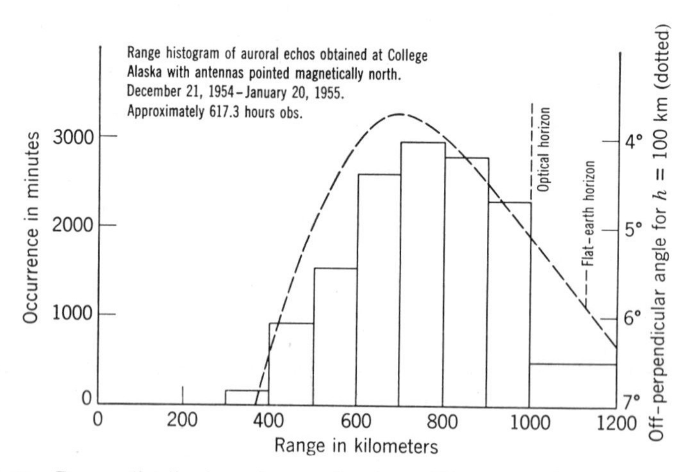

Fig. 19. Range distribution of auroral echoes (50 megacycles per second).[29]

velocities of the order of 600 meters per second. Not only are radar echoes subject to Doppler broadening (which causes the fading) but there is also a mean Doppler shift.[31] This is illustrated in Fig. 22 where the spectrum of the fading of auroral echoes is depicted. The spike at frequency F_0 represents the frequency transmitted by the radar. The upper and lower records were made simultaneously on two VHF frequencies differing by a factor of approximately 3. It can be seen that the Doppler shifts and Doppler spreads are proportional to trans-

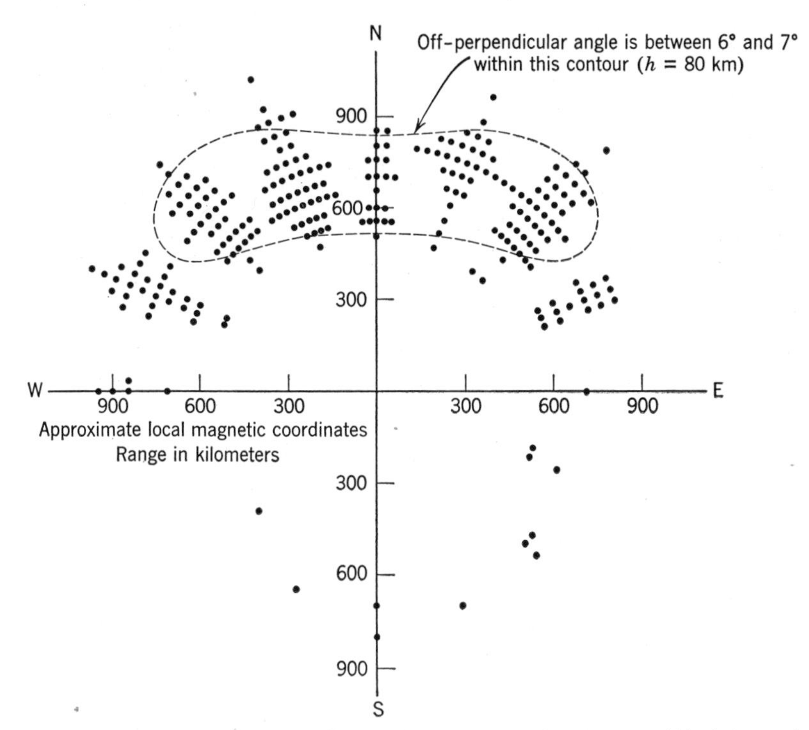

Fig. 20. Azimuth distribution of auroral echoes at Point Barrow (Alaska) on the polar side of the auroral zone (50 megacycles per second).[29]

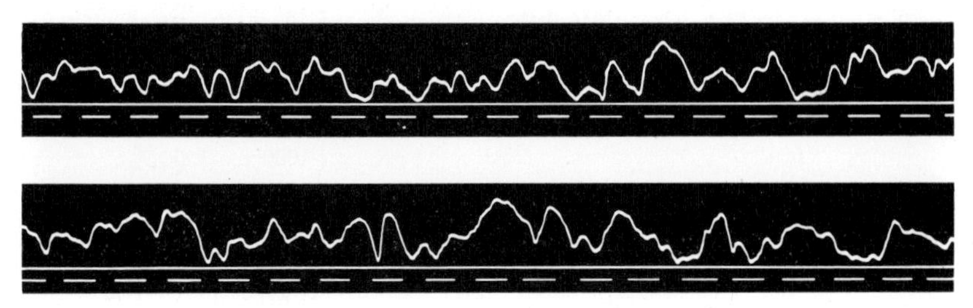

Fig. 21. Fading of auroral echoes (50 megacycles per second).[30]

mitted frequency, as would be expected if the phenomenon arose from some velocity associated with the ionosphere. What is surprising is that this velocity may be as high as 1000 meters per second.

It is also found that the variation in radar range of auroral echoes indicates high velocities of many hundreds of meters per second. Measurements, however, of velocities from the rate of change of radar

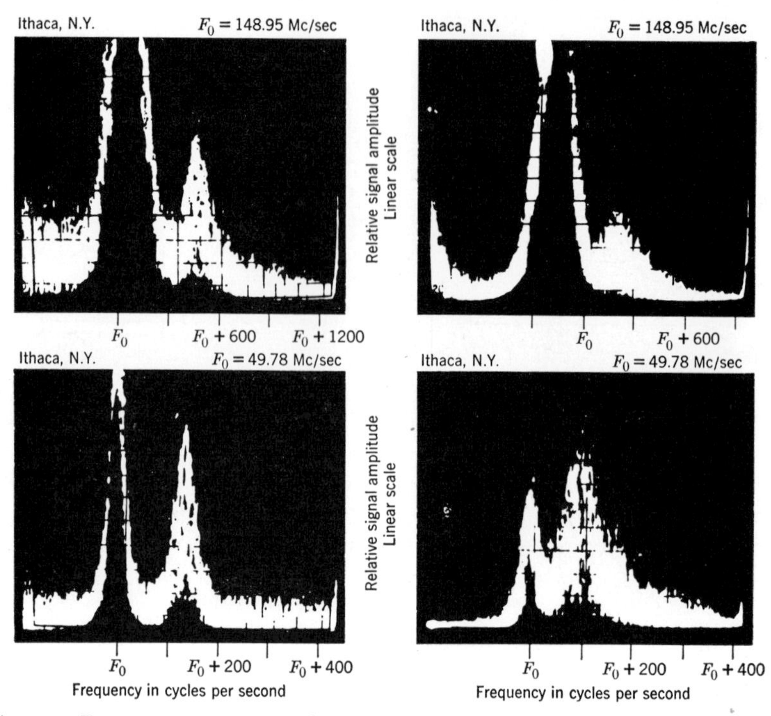

Fig. 22. Frequency spectrum of auroral echoes at 50 and 149 megacycles per second. F_0 is the transmitted frequency.

range and from the Doppler shift, although made simultaneously, do not give the same velocity and sometimes the signs of the two velocities are not even identical.

9. THE "SPORADIC *E*" PHENOMENON

During auroral activity an ordinary ionospheric recorder operating in the HF band may show scarcely any echoes as a result of the ionospheric attenuation associated with auroral ionization. This attenuation is not, however, always present and then records of the type shown in Fig. 23 may be obtained. It will be noticed that there is a strong return from a range of the order of the height of the *E* region. A phenomenon like this is usually described as indicating the presence of a "sporadic *E* region."

The phenomenon of sporadic *E* is not, however, confined to situations involving auroral activity. Fig. 24 shows two examples of

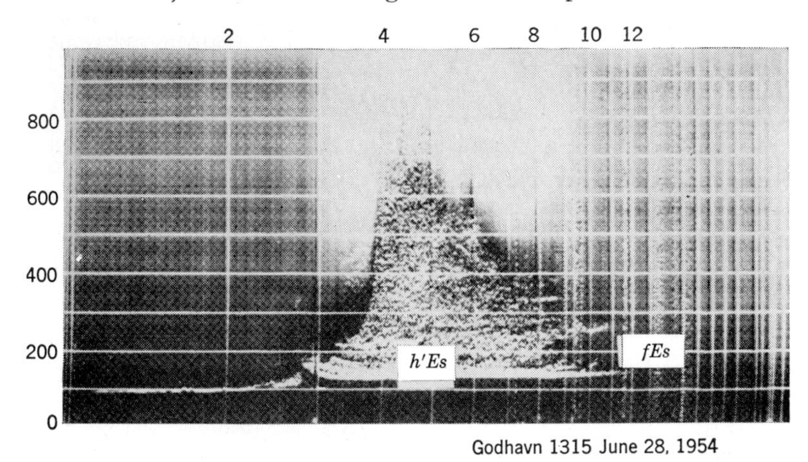

Fig. 23. Illustrating HF ionospheric record taken during auroral activity.[4]

sporadic *E* reflection of a more normal type than that shown in Fig. 23. In Fig. 24*a* the return from the level of the *E* region is only partial, and the echo from the *F* region can be seen through the sporadic *E* region. In Fig. 24*b*, however, the return is solely from the level of the *E* region. Although sporadic *E* echoes have now been studied for a quarter of a century, the cause of the phenomenon is still largely a mystery.

10. INTERPRETATION

Whether all nine of the scattering phenomena described before can be explained in terms of a single underlying phenomenon is not known at the present time. If there is one such phenomenon, it can scarcely be other than atmospheric turbulence. Atmospheric turbulence operating upon the regular ionization of the ionosphere can in principle explain the regular scattering phenomena, such as the fading of quiet ionospheric echoes (section 1). Atmospheric turbulence operating upon unusual increases in ionization can in principle explain some of the unusual scattering phenomena, such as the radar echoes from auroral ionization (section 8). There is also available the possibility of unusual increases in atmospheric turbulence, and of atmospheric turbulence extending to unusually great heights in the ionosphere.

Progress is being made with the explanation of the irregular phenomena associated with meteors (section 5). There is good reason to think that most of these phenomena will be explained in terms of the

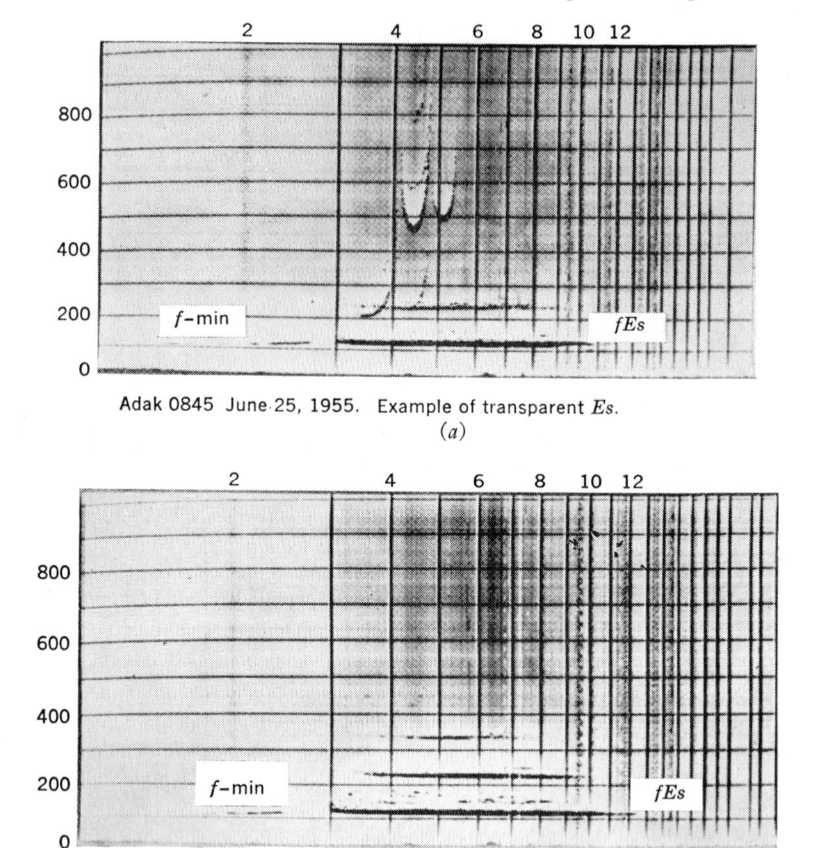

Adak 0845 June 25, 1955. Example of transparent *Es*.
(*a*)

Adak 0859 June 25, 1955. Example of blanketing *Es*.
(*b*)

Fig. 24. Sporadic reflection at the level of the *E* region: (*a*) transparent type; (*b*) blanketing type.[4]

usual Kolomogoroff-Heisenberg theory of turbulence, modified if necessary to allow for nonisotropy of the turbulence.[32, 24] The same is probably true of the background signal involved in VHF scatter communication (section 7), although the contribution of specular reflection from a large number of weak straight meteor trails is still under discussion.[24] It is a reasonable guess that VHF scatter communication involves the same phenomenon as that which produces HF echoes from below the *E* region (section 6) and consequently can be explained in the same way. This, however, has not so far been proved.

It was not at first suspected that the VHF echoes from auroral ionization (section 8) were connected with turbulence. The fact that auroral echoes are received only from low angles of elevation in roughly the northern quadrant can be explained if the scatterers are elongated along the earth's magnetic field. The length of the scatterers required to explain the observations quantitatively turns out to be of the order of a few tens of meters.[33] This size can most easily be explained if the irregularities are due to turbulence. The elongation of the electronic scatterers would arise from the effect of the earth's magnetic field in inhibiting motion of electrons across it while permitting motion parallel to it. On this viewpoint auroral echoes are simply due to scattering by electronic irregularities associated with turbulence, and the only function of the aurora would be to enhance the ionization-gradients involved and so render more intense the electronic irregularities produced by turbulence. If auroral echoes are explained in this way, then it becomes feasible to regard the non-auroral aspect-sensitive echoes described in section 3 as arising in the same way from ionization of other than auroral origin.

The most difficult observations of radio scattering in the ionosphere to explain are those concerned with the spread F phenomenon (section 2) and with star scintillation (section 4). Since the latter requires irregularities of electron density that are elongated along the earth's magnetic field,[34] the situation is similar in some respects to that described in sections 3 and 8. However, the higher one goes in the ionosphere, the more difficult it is to explain even the existence of turbulence. While the spread F and star scintillation phenomena are in all probability to be explained in terms of turbulence, no way has yet been found of using existing turbulence theories to provide a quantitative explanation.

REFERENCES

1. Appleton, E. V., and M. A. F. Barnett, On Some Direct Evidence for Downward Reflection of Electric Rays, *Proc. Roy. Soc. (London)*, **A 109**, 621 (1925).
2. Breit, G., and M. A. Tuve, A Test of the Existence of the Conducting Layer, *Phys. Rev.*, **28**, 554 (1926).
3. Mitra, S. K., *The Upper Atmosphere*, 2nd ed., Asiatic Society, Calcutta (1952).
4. Wright, J. W., and R. W. Knecht, *Atlas of Ionograms*, National Bureau of Standards, Washington, D. C., prepared for the International Geophysical Year, 1957.

5. McNicol, R. W. E., The Fading of Radio Waves of Medium and High Frequencies, *Proc. Inst. Elec. Engrs. (London)*, **96**, Part III, 517 (1949).
6. Reber, G., Spread *F* over Washington, *J. Geophys. Research*, **59**, 445 (1954); World-Wide Spread *F*, *J. Geophys. Research*, **61**, 157 (1956).
7. Bibl, K., E. Harnischmacher, and K. Rawer, Some Observations of Ionospheric Movements, *The Physics of the Ionosphere*, Physical Society of London, p. 113 (1955).
8. Eckersley, T. L., Recombination and Spread Echoes from the Ionosphere, *Proc. Phys. Soc. (London)*, **B, 66**, 1025 (1953).
9. Peterson, A. M., O. G. Villard, Jr., R. L. Leadabrand, and P. B. Gallagher, Regularly-Observable Aspect-Sensitive Radio Reflections from Ionization Aligned with the Earth's Magnetic Field and Located Within the Ionospheric Layers at Middle Latitudes, *J. Geophys. Research*, **60**, 497 (1955).
10. Ryle, M., and A. Hewish, The Effects of the Terrestrial Ionosphere on the Radio Waves from Discrete Sources in the Galaxy, *Monthly Notices Royal Astronomical Society*, **110**, 381 (1950).
11. Little, C. G., and A. Maxwell, Fluctuations in the Intensity of Radio Waves from Galactic Sources, *Phil. Mag.*, **42**, 267 (1951).
12. Seeger, C. L., Some Observations of the Variable 205 Mc/sec Radiation of Cygnus A, *J. Geophys. Research*, **56**, 239 (1951).
13. Wild, J. P., and J. A. Roberts, The Spectrum of Radio-star Scintillations and the Nature of Irregularities in the Ionosphere, *J. Atmospheric and Terrest. Phys.*, **8**, 55 (1956).
14. Greenhow, J. S., Characteristics of Radio Echoes from Meteor Trails: III. The Behaviour of Electron Trails after Formation, *Proc. Phys. Soc. (London)*, **B, 65**, 169 (1952).
15. ———, The Fluctuation of Radio Echoes from Meteor Trails, *Phil. Mag.*, **41**, 682 (1950).
16. ———, Systematic Wind Measurements at Altitudes 80–100 km Using Radio Echoes from Meteor Trails, *Phil. Mag.*, **45**, 471 (1954).
17. Liller, W., and F. L. Whipple, High-altitude Winds by Meteor-train Photography, *Rocket Exploration of the Upper Atmosphere*, ed. by R. L. F. Boyd and M. J. Seaton, supplement to *J. Atmospheric and Terrest. Phys.*, **2**, Pergamon Press, London, p. 112 (1954).
18. de Jager, C., The Spectrum of Turbulence in the Earth's Upper Atmosphere, *Mem. Soc. roy. sci. Liége*, **12**, 223 (1952).
19. Dieminger, W., Short-Wave Echoes from the Lower Ionosphere, *The Physics of the Ionosphere*, The Physical Society of London, p. 53 (1955).
20. Gardner, F. F., and J. L. Pawsey, Study of the Ionospheric D-region Using Partial Reflections, *J. Atmospheric and Terrest. Phys.*, **3**, 321 (1953).
21. Bailey, D. K., R. Bateman, L. V. Berkner, H. J. Booker, G. F. Montgomery, E. M. Purcell, W. W. Salisbury, and J. B. Wiesner, A New Kind of Radio Propagation at Very High Frequencies Observable over Long Distances, *Phys. Rev.*, **86**, 141 (1952).
22. Bailey, D. K., R. Bateman, and R. C. Kirby, Radio Transmission at VHF by Scattering and Other Processes in the Lower Ionosphere, *Proc. I.R.E.*, **43**, 1181 (1955).
23. Eshleman, V. R., and L. A. Manning, Radio Communication by Scattering from Meteoric Ionization, *Proc. I.R.E.*, **42**, 530 (1954).

24. Booker, H. G., and R. Cohen, A Theory of Long-Duration Meteor-Echoes Based on Atmospheric Turbulence with Experimental Confirmation, *J. Geophys. Research*, **61**, 707 (1956).
25. Dyce, R., VHF Auroral and Sporadic-E Propagation from Cedar Rapids, to Ithaca, New York, *Trans. I.R.E.*, **AP-3** (No. 2), 77 (1955).
26. Bowles, K. L., Some Recent Experiments with VHF Radio Echoes from Auroras and Their Possible Significance in the Theory of Magnetic Storms and Auroras, Cornell University Ph.D. thesis, 1955.
27. Booker, H. G., C. W. Gartlein, and B. Nichols, Interpretation of Radio Reflections from the Aurora, *J. Geophys. Research*, **60**, 1 (1955).
28. Dyce, R., More About VHF Auroral Propagation, *QST*, **39**, 11 (1955).
29. ———, Auroral Echoes Observed North of the Auroral Zone on 51.9 Mc/sec, *J. Geophys. Research*, **60**, 317 (1955).
30. Bowles, K., The Fading Rate of Ionospheric Reflections from the Aurora Borealis at 50 Mc/sec, *J. Geophys. Research*, **57**, 191 (1952).
31. ———, Doppler Shifted Radio Echoes from Aurora, *J. Geophys. Research*, **59**, 553 (1954).
32. Booker, H. J., Phenomena of Irregular Scattering in the Ionosphere, *J. Geophys. Research*, **61**, 347 (1956).
33. ———, A Theory of Scattering by Nonisotropic Irregularities with Application to Radar Reflections from the Aurora, *J. Atmospheric and Terrest. Phys.*, **8**, 204 (1956).
34. Spencer, M., The Shape of Irregularities in the Upper Ionosphere, *Proc. Phys. Soc. (London)*, **B**, **68**, 493 (1955)
35. McKinley, D. W. R., The Meteoric Head Echo, *Meteors*, ed. by T. R. Kaiser, supplement to *J. Atmospheric and Terrest. Phys.*, **2**, Pergamon Press, London and New York, p. 65 (1955).